GREAT DISCOVERIES IN SCIENCE

DNA and RNA

by Jenny J. Chen

New York

Published in 2017 by Cavendish Square Publishing, LLC
243 5th Avenue, Suite 136, New York, NY 10016

First Edition

CPSIA Compliance Information: Batch #CS16CSQ

All websites were available and accurate when this book was sent to press.

Library of Congress Cataloging-in-Publication Data

Names: Chen, Jenny J.
Title: DNA and RNA / Jenny J. Chen.
Description: New York : Cavendish Square Publishing, [2017] |
Series: Great discoveries in science | Includes bibliographical references and index.
Identifiers: LCCN 2016006889 (print) | LCCN 2016012678 (ebook) |
ISBN 9781502619556 (library bound) | ISBN 9781502619563 (ebook)
Subjects: LCSH: DNA. | RNA. | Genetics.
Classification: LCC QP624 .C54 2017 (print) |
LCC QP624 (ebook) | D DC
572.8/6--dc23
LC record available at http://lccn.loc.gov/2016006889

Editorial Director: David McNamara
Editor: Caitlyn Paley
Copy Editor: Michele Suchomel-Casey
Art Director: Jeffrey Talbot
Designer: Joseph Macri
Production Assistant: Karol Szymczuk
Photo Research: J8 Media

The photographs in this book are used by permission and through the courtesy of: Andrii Vodolazhskyi/Shutterstock.com, cover; SEBASTIAN KAULITZKI/Science Photo Library/Getty Images, 4; Jastrow/Ludovisi Collection/File:Aristotle Altemps Inv8575.jpg/Wikimedia Commons, 8; The Print Collector/Print Collector/Getty Images, 11; Capreola/Shutterstock.com, 15; wellcome trust/File:Johann Friedrich Miescher. Photograph. Wellcome V0026860.jpg/Wikimedia Commons, 17; NoPainNoGain/Shutterstock.com, 21; Science Source/Getty Images, 26; Unknown/File:Levene.jpg/Wikimedia Commons, 32; JWSchmidt/File:Tetranucleotide.JPG/Wikimedia Commons, 33; Horst Tappe/Getty Images, 36; Time Life Pictures/Mansell/The LIFE Picture Collection/Getty Images, 41; George Silk/The LIFE Picture Collection/Getty Images, 46; DANIEL MORDZINSKI/AFP/Getty Images, 48; © Everett Collection Historical/Alamy Stock Photo, 44; AP Photo/Alastair Grant, 51; NLM/Science Source/Getty Images, 53; Randy Leffingwell/Los Angeles Times via Getty Images, 61; Science Source/Getty Images, 64; E. O. Hoppe/Mansell/The LIFE Picture Collection/Getty Images, 66; Don Cravens/The LIFE Images Collection/Getty Images, 71; © Everett Collection Historical/Alamy Stock Photo, 75; LAGUNA DESIGN/Science Photo Library/Getty Images, 82; Mondadori Portfolio/Getty Images, 85; Gio.tto/Shutterstock.com, 88; Jarrod Erbe/Shutterstock.com, 91; © ZUMA Press, Inc./Alamy Stock Photo, 95; Keith Weller, U.S. Department of Agriculture/File:Genetically modified corn.jpg/Wikimedia Commons, 101; Prachatai/salmon aquabounty/flickr, 103.

Printed in the United States of America

Contents

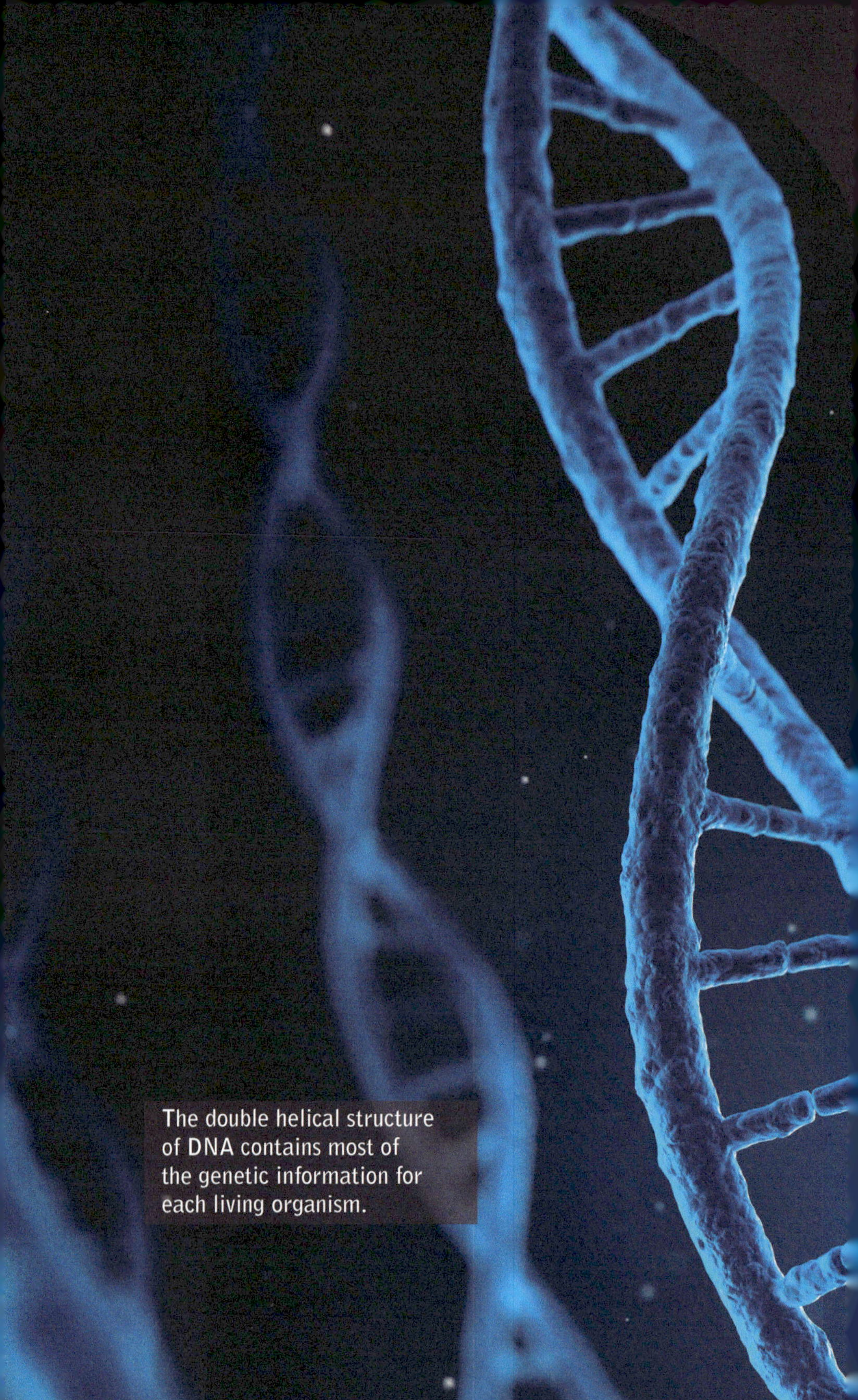

The double helical structure of DNA contains most of the genetic information for each living organism.

Introduction

You've probably heard about the discovery of **DNA** before. You've probably heard of Watson and Crick, the two scientists working in England who solved the mystery of the double **helix**. You might imagine two scientists sitting in a musty basement, scratching out boring mathematical problems to the rhythm of a clock on the wall. But the story of DNA is much more interesting than that. It involved some of the most brilliant people of the century. Yet their stories are so human, at times you might feel that you recognize them in people you know and even in yourself. Perhaps you will be intrigued by Linus Pauling, a scientist so famous at the time that he might as well have been a superstar. Perhaps you will come to know and love Rosalind Franklin, the largely misunderstood brilliant physicist who is often overlooked in the story of DNA. Maybe you'll be touched by the story of Maurice Wilkins, the physicist so disgusted by his work with the atom bomb that he vowed to study the science of life and never again the science of death.

You will be struck by the tale of something so incredible and mind-blowing yet dear and close to our lives. After all, the race to discover DNA was unlike anything in history. It would finally give scientists a clue as to what makes up the fabric of life. It asked and answered the question of who we really are and what we are made of. The story of DNA is everywhere around us. It's in the food that we eat, in our restaurants, in the supermarkets, and in our farms. It has helped us save lives and understand diseases like sickle-cell anemia. From cloning to **eugenics**, DNA is at the heart of it all.

Our story begins in ancient Greece, a time when people still thought the sun revolved around Earth. They took some early stabs at how inheritance worked—some people thought that only men passed on their genetic information. Others thought that people started out fully formed out of nowhere.

We then travel to Germany, where a humble monk (who didn't really plan to be a monk but became one so he could get the education) named Gregor Mendel carefully studied pea plants to see how they passed on their traits. Mendel was able to start answering the questions: What traits from each parent get passed on? Why do some traits seem to skip a generation?

Elsewhere in a gloomy isolated castle in Tubingen, Germany, we will then meet Friedrich Miescher, who isolated a substance nobody had ever seen before. That substance would later turn out to contain the code for living things, but he didn't know it at the time. We then move through a slew of other German scientists, including Walther Flemming and Oskar Hertwig, who discovered how cells divide and replicate themselves.

Our story will then take us to Great Britain, home to scientists like Sir William Lawrence Bragg and William Astbury, who would pioneer the use of X-rays to look at tiny particles. By this time, interest in DNA had spread all over the world. Scientists like Phoebus Levene in Russia would begin breaking down the components of DNA. Erwin Schrodinger in Austria would tackle the question of how DNA was structured through the lens of physics, and scientists in the United States would prove definitively through elegantly designed experiments that DNA was the carrier of genetic material.

But when it came to the final discovery, we return once again to Great Britain, where the race to find the structure of DNA finally came to a head with scientists James D. Watson, Francis Crick, Rosalind Franklin, and Maurice Wilkins. They were opposed only by Linus Pauling in America, regarded as the father of modern biology.

We'll then move from the discovery of DNA to the discovery of **RNA**, which involves scientists such as Marshall Nirenberg and Johann H. Mathaei, who were able to figure out how DNA created the building blocks of life.

Finally, we'll arrive in the present-day United States, where the discovery of DNA has allowed us to do previously unbelievable things, including mapping out the instructions for coding a human being. We will talk about how our knowledge of DNA has allowed us to change the **genes** of the food we eat, the genes of **viruses** and **bacteria** that surround us, and the genes of the animals that live alongside us. We'll also talk about the scary reality of how close we are to changing human genes—and possibly changing the human race forever.

The tale of DNA is a special one in science. It asks the questions that humans have wondered since the beginning of time: Who are we? Why are we here? Perhaps this is why the question of DNA captivated so many scientists over the last century.

Aristotle was an ancient Greek philosopher and scientist. Rumor has it that he wrote more than two hundred essays, but only thirty-one exist today.

CHAPTER 1

The Problem of Inheritance

Take a look at yourself in the mirror. You probably see characteristics from both your mother and father in your face. Your mother's nose, your father's chin. Maybe even your aunt's hair, which no one else in the family has except for you two. All these traits are what make you uniquely you. But how exactly did you get them? And why did you get your father's chin, but not his eyes?

Since the beginning of time, people have asked this question. Ancient farmers knew that when you bred two animals together, the resulting offspring would have the combined characteristics of both the mother and father, but they didn't know how. Great thinkers began to develop theories of inheritance, starting with the ancient Greeks.

TINY HUMANS and SEEDS: HOW PEOPLE FIRST THOUGHT INHERITANCE WORKED

The ancient Greeks had a theory they called "preformation." The theory stated that humans started off as fully formed, miniature humans, or "preformed," in the mother's womb. The theory also stated that the man's semen contained the soul, or the “life essence,” necessary to make a human being and the mother was just the incubator that brought the “life essence” to the world. In a book called *Timaeus*, the Greek philosopher Plato wrote that men “sow in the womb, as in a field.” They also believed that semen was created in the man’s brain and traveled down to the testicles.

But there were large holes in this theory, the biggest one being something that you already know—children look like both their mother and father, not just their father alone. A few years later, Aristotle and a few other Greek philosophers suggested another theory called **epigenetics**. The word “epigenetics” comes from a Greek word that means “extra growth.” The theory of epigenetics stated that humans were not fully formed but created first as undistinguishable masses before they developed into a distinct human being. Aristotle came up with this theory after observing chicks. He noted that the early egg was not an already formed version of a miniature chicken but only started developing chicken-like characteristics (a beating heart, a beak, feathers) after it continued growing. Aristotle theorized that the semen from the father came together with the menstrual blood in the mother to create an unformed mass, which would later grow into a formed fetus.

The debate between the preformationists and the epigeneticists raged for years. Some thought that the idea of a tiny man in a mother’s body was ridiculous, while others questioned how an unformed mass could miraculously become a formed human being.

In 1859, Charles Darwin added his own thoughts. He was part of a group of scientists who believed in the theory of **pangenesis**. The prefix “pan” means “all,” and Darwin believed that information from all the cells in both parents was responsible. He believed that all the cells of the body carried tiny seeds called gemmules and that all the gemmules carried a bit of information from that organ to be passed on as an exact replica to offspring. He did not believe that the gemmules

Charles Darwin was an English naturalist and geologist best known for developing a theory of evolution.

carried the entire blueprint for the entire organism. Many people criticized the pangenesis theory, however, because according to the theory, if a parent acquired a trait through life that he or she wasn't born with—like more muscles or scars—that trait could be passed on to the children. Everyone knew that this didn't happen. Darwin's half-cousin Frances Galton tried to see if Darwin's proposed gemmules could be transferred through blood. He conducted several experiments from 1869 to 1871 that involved the transfusion of blood between different breeds of rabbits. He mated a rabbit with transfused blood with a rabbit of the same breed without transfused blood and examined the characteristics of the offspring. Galton did not find any evidence of the different breed of rabbit in the offspring, which pretty much disproved the idea that there were gemmules floating around in our cells.

OF PEAS and MEN: HOW a MONK FIGURED OUT the FIRST LAWS of INHERITANCE

At the same time that Darwin was writing about gemmules, another scientist was coming up with a radically different idea of inheritance and genetics. He wasn't a scientist by traditional definitions—he was actually a monk.

Gregor Mendel was born Johan Mendel in the Austrian Empire (present-day Czech Republic). His family lived as poor farmers, but Mendel had greater aspirations for his education. He enjoyed learning and studied philosophy and physics at the University of Olomouc in Moravia. He and his family struggled to fund his education, and his younger sister Theresia even gave him part of her wedding dowry.

In order to finish his education, Mendel became a friar in the Augustine order. At the time, the Augustinians believed in giving free education to those who served in their religious order. They funded his education at the University of Olomouc, where he studied under Johann Karl Nestler. Nestler researched how traits get passed down in animals and plants, which was called the science of heredity. Inspired by this work, Mendel continued to research heredity after he joined St. Thomas's Abbey as a priest.

Mendel began doing his own experiments on inheritance. He worked with pea plants because they were relatively simple (unlike humans or animals) and they could either self-pollinate (which means producing offspring using only their own genes) or cross-pollinate (which means combining genes with another pea plant to create baby pea plants). They were also widely available. Mendel created a list of seven different characteristics of the pea plants, including flower color (purple or white), flower position (axial or terminal), stem length (long or short), seed shape (round or wrinkled), seed color (yellow or green), pod shape (inflated or constricted), and pod color (green or yellow). Then, he carefully bred the pea plants and observed the characteristics of the offspring, one at a time. It was a very time-consuming process. In his most famous experiment, he bred a yellow-seeded pea plant with a green-seeded pea plant. The resulting offspring pea plants all had yellow seeds. He then bred all the offspring yellow-seeded plants with each other. Some of those third-generation pea plants ended up having green seeds even though none of their parents had green seeds! From this experiment and many others, Mendel came up with what later became known as Mendel's laws.

Mendel's Laws of Genetic Dominance

As a result of his experiments, Mendel theorized that some traits were "dominant" over others. In the case of the pea plants, for example, the yellow-seed one was dominant, so using principles of mathematics, Mendel theorized that each parent provided two pieces of genetic information (what we now call **alleles**). These alleles can be dominant or recessive. Recessive traits will show themselves only if there is no dominant trait allele. In the case of the pea plant experiment, the yellow-seeded pea plant carried two dominant yellow-seed alleles. The green-seeded pea plant carried two recessive green-seed alleles. When the two plants breed and each one contributes one allele to the offspring, each offspring ends up with a yellow-seed allele and a green-seed allele. Because the yellow-seed allele is dominant, all the offspring display the yellow-seed trait. But in reality, they each carry the hidden green-seeded allele. When these offspring with mixed alleles are bred together, they have a 25 percent chance of producing an offspring that carries only the recessive green-seeded alleles and, therefore, express the green-seed trait, even if both parent pea plants are yellow-seeded. These traits are either/or, which means that a pea plant is either smooth or wrinkled, not in between (there can be some exceptions for more complicated traits like hair color or skin color). This finding contradicted what a lot of scientists believed at the time, which was that parent traits "blended" in their children.

Mendel also found that traits were passed on independently from each other, which means that whether a pea plant is yellow or green has nothing to do with whether a pea plant produces smooth or wrinkled peas. In humans, this means you can have your father's eyes and your mother's nose at the same time.

Mendel never received much recognition for his groundbreaking experiments. However, years after his death, other scientists would go on to confirm what he found. In the 1890s, Dutch botanist Hugo de Vries replicated Mendel's experiments with various plant species. Many argue that de Vries was unaware of the significance of his findings until he read Mendel's work. In 1892, Carl Cowens, a German botanist, also independently confirmed Mendel's work. After that, the scientific community became extremely excited about Mendel's laws and began building upon them. The modern age of molecular biology was about to begin.

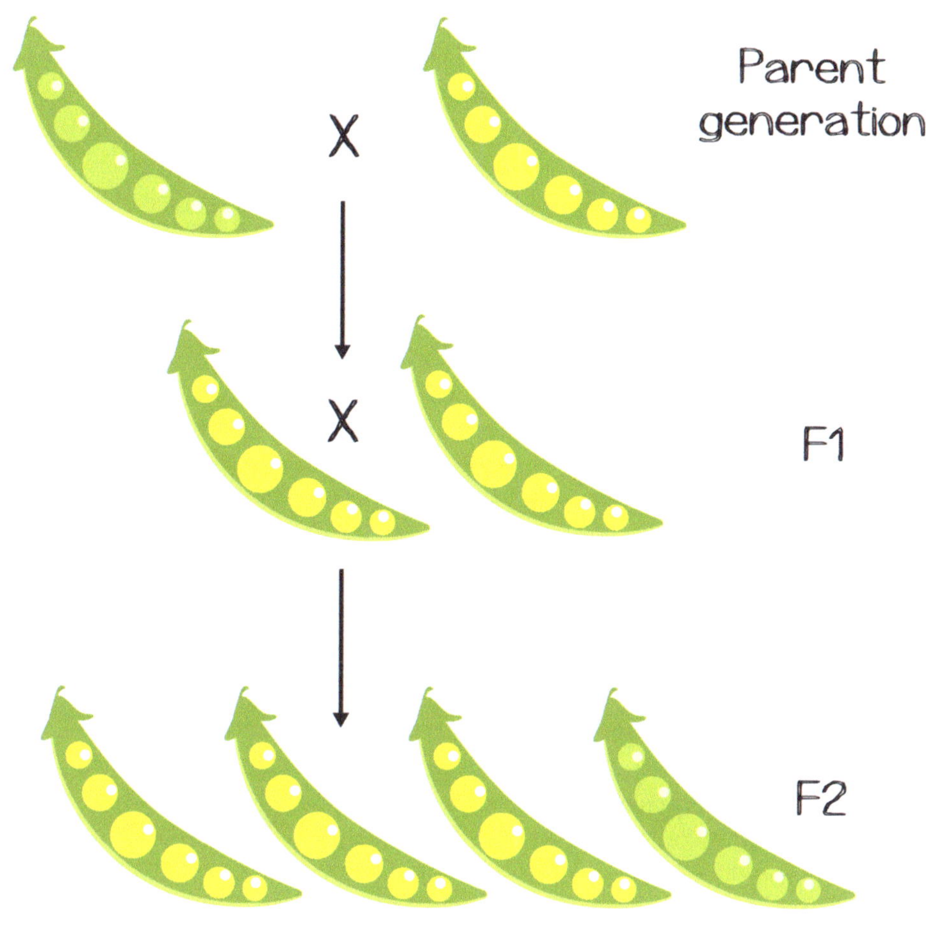

Originally, Augustinian monk Gregor Mendel wanted to study genetics by breeding mice, but his supervisors did not like the idea.

The DISCOVERY of NUCLEIN

Friedrich Miescher was born in Basel, Switzerland, on August 13, 1844, to a family of scientists. One of the biggest influences of his life was Wilhelm His, his uncle, who became famous for his research on cell and tissue development for embryos. Miescher's father wanted him to study to be a doctor, so at seventeen years old, Miescher started medical school. However, his heart wasn't in it. He also had trouble hearing his patients because he had suffered from a serious ear infection that had permanently damaged his hearing when he was a child. Seeing how unhappy Miescher was as a doctor, his uncle suggested that he pursue a career in research instead and pointed him in the direction of organic chemistry—the study of everything and anything with the element carbon.

Miescher worked under Felix Hoppe-Seyler, a prominent physiologist and chemist who helped develop the fields of biochemistry, physiological chemistry, and molecular biology. Like Miescher, Hoppe-Seyler had originally studied to be a doctor but found research more interesting. When Miescher met him, Hoppe-Seyler was busy studying human fluids like blood, pus, and milk at the isolated Tubingen Castle in the heart of Germany. Miescher worked alongside Hoppe-Seyler for a while and developed an interest in figuring out exactly how cells worked by studying lymphocytes, small white blood cells important for the immune system. However, it was extremely difficult to obtain lymphocytes since they had to be taken from patients' lymph nodes. Hoppe-Seyler suggested that his student study **leucocytes** instead, since they could easily be obtained from the pus on patients' bandages.

At first, Miescher worked on isolating the **proteins** of leucocytes since at the time, people believed that proteins were the most important part of how cells functioned. Miescher had very limited access to scientific tools and technology, however, and so had to get creative. He first tried washing old pus-filled bandages with a solution of sodium chloride (NaCl), but the

Although largely forgotten by history, German scientist Friedrich Miescher was one of the first people to discover the DNA molecule.

solution made the cells swell up and turn slimy, which made them impossible to work with. The same thing happened when he added alkali solutions. Finally, Miescher was able to isolate the leucocytes by washing the cells in a diluted solution of sodium sulfate, which successfully separated the leucocytes from the rest of the cells.

But Miescher was still really interested in the sliminess that oozed out of the cells when an alkali solution was added to the leucocytes. After experimenting with the slimy solution some more, Miescher realized that even though it looked like a protein, it was actually not a protein. It couldn't be digested by the protein enzyme **protease** pepsin. It didn't contain sulfur, which was an ingredient in all other proteins. But if it wasn't a protein, what was it?

By burning the substance and observing the reaction, Miescher found that it contained carbon, hydrogen, oxygen, nitrogen, and large amounts of phosphorous. Through further experimentation, he also found the substance in other cells other than leucocytes and so guessed that it was probably really important to the way cells functioned. Because it was found in the **nucleus** of cells, Miescher called the substance "nuclein." He also suspected the substance carried genetic material but wasn't able to prove it.

The more he discovered about the substance, the more convinced he was that he had made a new discovery, something scientists hadn't known before, and he hoped that other scientists would use what he had found to continue digging into the secrets of the cell. "Knowledge of the relationship between nuclear substances, proteins and their closest conversion products will gradually help to lift the veil which still utterly conceals the inner processes of cell growth," Miescher wrote in 1871.

It took a while for the scientific community to understand the significance of Miescher's findings, but his work did pique the interest of a small group of scientists. Albrecht Kossel, a

German biochemist, did a series of experiments from 1885 to 1901 to isolate and figure out the five organic **molecules** that make up DNA and RNA—adenine, thymine, cytosine, guanine, and uracil. In 1881, the German botanist Eduard Zacharias was able to link **chromosomes** to DNA, which further proved that DNA carried genetic material. But it wasn't until the mid-1940s when the full significance of nuclein became known.

HOW CELLS PASS on THEIR GENETIC MATERIAL

Around the same time that Miescher discovered nuclein, another pair of German scientists, Walther Flemming and Oskar Hertwig, were watching cells divide. They figured out how chromosomes, tiny threadlike substances that contain DNA, replicated and divided to form new cells. These breakthrough discoveries solved the mystery of how traits from both the mother and the father were passed on to offspring. They also disproved previous theories of preformation and epigenetics.

Walther Flemming and Mitosis

Flemming was born in Sachsenberg, Germany. His father was the famous psychiatrist and neuroscientist Carl Friedrich Flemming. Flemming studied medicine at the University of Gottingen. After getting his degree, Flemming went on to work at a clinic in Rostock, where he learned the basics of science and conducting experiments from his mentor, Franz Eilhard Schulze. In 1872, Flemming was asked to join the German University of Prague (in present-day Czech Republic). This is where he began investigating cell division. Unfortunately, the political climate in Prague was not kind to Germans, and

Flemming eventually returned to Germany to teach at the University of Kiel.

While he was teaching anatomy at the University of Kiel, Flemming used aniline dyes to watch cells as they divided. Aniline is an oily, poisonous, black substance taken from the indigo plant. The entire research process was difficult because there were limited microscopes available to all the researchers. Flemming himself said that his lab "[lacked] space and [lacked] material."

Flemming called the cell division process that he discovered **mitosis**. The process was based off of a substance called chromosomes. Chromosomes look like tiny, tiny worms. Humans have forty-six chromosomes. When chromosomes replicate, two copies of the same chromosome form an "x." The x is held together with something called a centromere. Each of the individual copies in the "x" are called **chromatids**.

Mitosis is the process that body cells use to replicate themselves. This process is important as living organisms grow (and therefore gain more cells), or when a part of the organism gets hurt and requires new cells for repair. Mitosis happens in the nucleus of the cell, where all the genetic information is stored. There are four main phases to mitosis: prophase, metaphase, anaphase, and telophase.

Before mitosis even begins, the cell makes a copy of its chromosomes. So if a human cell starts off with forty-six chromosomes, it will have ninety-two by the time mitosis begins. In the first phase of mitosis, prophase, the chromosomes thicken and get ready to divide. In the second phase of mitosis, metaphase, those ninety-two chromosomes line up in the center of the cell. Then the chromosomes separate to opposite sides of the cell in a process call anaphase. Finally, in telophase, a new nucleus forms around each group of chromosomes. The cell then splits into two separate cells. This is how one parent cell can produce two cells, which are called daughter cells.

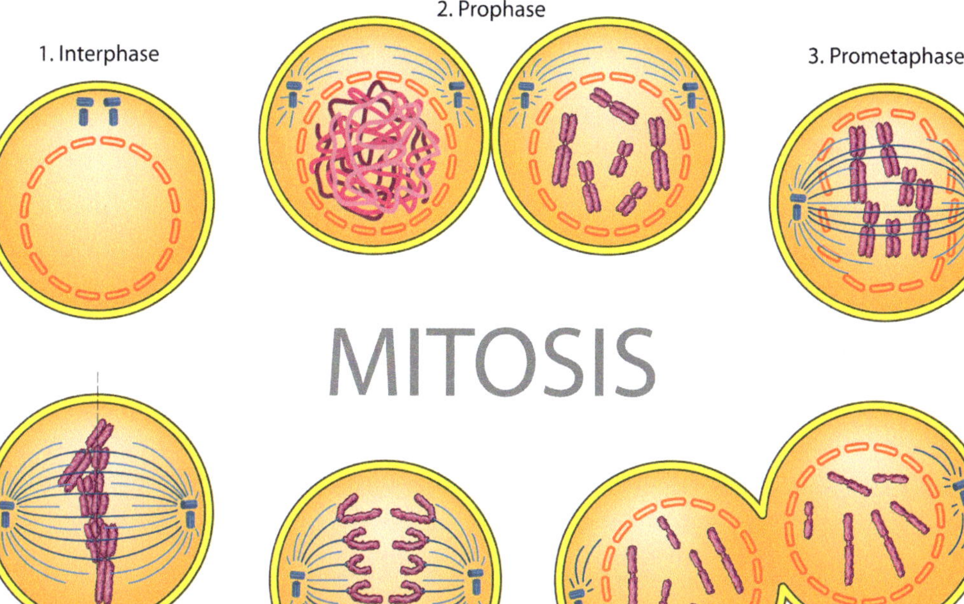

Cells go through a process called mitosis to pass genetic material on to new cells.

The process of mitosis only occurs for body cells, like your skin cells or heart cells. Later on, Flemming's contemporary, Oskar Hertwig, discovered another replication process by observing the eggs of sea urchins.

Oskar Hertwig and Meiosis

Oskar Hertwig was born on April 21, 1849, in Hessen, Germany. He and his brother went to college together at the University of Jena. They studied with Ernst Haeckel, a well-respected German biologist. Haeckel inspired both brothers to study embryology—the study of unborn living beings.

Eventually, Haeckel convinced Oskar Hertwig to switch from studying chemistry to medicine. Hertwig became a professor of anatomy at the University of Jena in 1901, and his brother became a professor of zoology at the University of Munich.

In 1876, Hertwig began investigating the process of fertilization in sea urchin eggs. Sea urchin eggs are transparent, which makes it easy for scientists to observe what's going on in inside. Because of this, Hertwig was able to watch the movement of chromosomes as egg cells divided. He was also able to watch as sperm cells fused with egg cells and combined genetic information. In addition, he was able to prove that an egg needs to fuse with only one sperm cell to fertilize. Once a sperm cell has reached the egg cell, the egg forms a thick outside layer that blocks any other sperm cells from coming through.

Through his experiments, Hertwig discovered **meiosis**. Meiosis is how sperm and egg cells replicate. Meiosis is similar to mitosis, but the goal of the process is not to make cells identical to the original cell. In meiosis, the resulting daughter cells have only half the chromosomes as the original cell.

Meiosis begins with the chromosomes making a copy of themselves, just like in mitosis. Just like in mitosis, we begin with ninety-two chromosomes. Also like mitosis, the chromosomes thicken in prophase to prepare to divide. However, when the chromosomes line up in the middle of the cell in metaphase, they line up in pairs with their homologous pair. A homologous pair is a chromosome that codes for pretty much the same genes but comes from two different parents. For example, let's take the gene that codes for hair color. Let's say your father has black hair and your mother has red hair. You would have the chromosomes that code for hair color from both your father and your mother. The one from your father would pair up with the one from your mother in metaphase. The pairs exchange a bit of genetic information with each other in a process called chromosomal crossover (this helps ensure

The Discovery of Chromosomes

In 1879, the German biologist Walther Flemming discovered tiny threadlike substances in the nucleus. These threadlike substances actually contained bundles of DNA wrapped around a type of protein called histone. He used a dark dye called aniline dye on salamander embryos so he could watch the threadlike substances divide when the cell made copies of itself. When the researchers zoomed in on the cell to watch it divide and replicate genetic information, they saw that these threadlike substances would pair up before the cell split and then divide so that a bit of each thread would be in the new pairs. All the resulting baby cells would also get a copy. Flemming called this substance **chromatin** because it could easily be stained with color (the prefix "chroma" comes from the Greek word for color). Some scientists, including German researcher Oskar Hertwig, found that chromatin contained DNA, which made them guess that it carried genetic material. However, chromatin also carried a lot of protein, so many scientists disagreed with Flemming and said that proteins carried genetic information.

that there is a lot of variety in different species as genes get passed on).

After the homologous chromosomes line up in the center of the cell, the cell goes through anaphase, just like in mitosis. The pairs of chromosomes get separated to opposite ends of the cell. A nucleus forms around each group of chromosomes, and two new cells are formed. But the process isn't done yet. Each of the daughter cells goes through division *again*, in a process called meiosis II. Keep in mind that each of the daughter cells now has forty-six chromosomes and they do not replicate again. In meiosis II, those forty-six chromosomes line up in the center again (this time in single file). In the second anaphase, the centromeres that hold the "x" together unravel and the two pieces of the chromosome separate. Those pieces are now called chromatids. Each chromatid gets pulled to opposite ends of the cell. A nucleus forms around each group of chromatids. In the end, meiosis produces a total of four daughter cells. This is different from mitosis, which produces just two daughter cells. By meiosis, each of the daughter cells has only half of the DNA of the original cells. This is important because each sperm and egg cell contributes only half of the genetic material to the egg cell. In 1883, Belgian zoologist Edouard Van Beneden confirmed Hertwig's findings when he studied roundworm eggs.

FROM GEMMULES to GENES

Science has come a long way since ancient times. We started out believing that only fathers were responsible for passing on their genetic information. We then found the actual substance that carries our genes—nuclein, which would eventually be called DNA. We also figured out how our cells divide and pass on genetic information. At this point, the mysteries of inheritance were a little bit less mysterious.

But there were still some major holes in our understanding of inheritance. For one, many scientists were still not convinced that nuclein (or DNA) held genetic material. A lot of them still believed that proteins held genetic material. This is partly because the substance of DNA itself was still a huge mystery—no one knew what it actually looked like or how it behaved. But it wouldn't take long for DNA to take center stage in figuring out how human life was created.

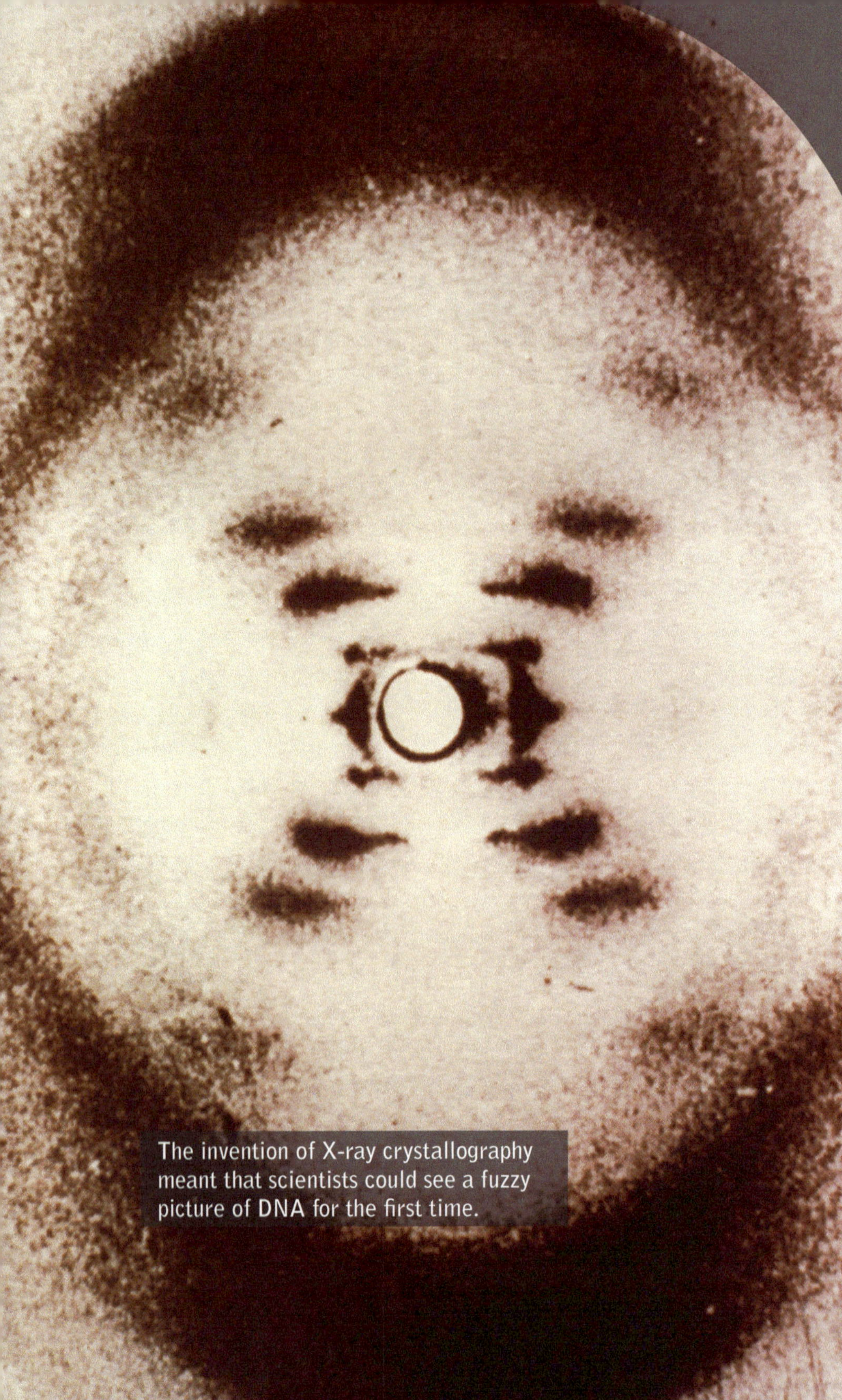

The invention of X-ray crystallography meant that scientists could see a fuzzy picture of DNA for the first time.

CHAPTER 2

The Science of DNA and RNA

By the late 19th century, scientists knew that genetic material was contained in the cell. Friedrich Miescher had also discovered the DNA substance, which he called "nuclein" because it was found in the nucleus of the cell. But even though Miescher thought that DNA might contain genetic information, no one knew for sure. Many scientists thought that proteins contained genetic material. In fact, many of them thought that DNA wasn't as complex as the other molecules in the body and they called it "the stupid protein." It wasn't until the Avery-MacLeod-McCarty experiments that scientists began to gather proof that DNA contained genetic information. After this remarkable experiment, scientists began debating the structure of DNA. These experiments laid the groundwork for our modern understanding of both DNA and RNA.

OSWALD AVERY and the AVERY-MACLEOD-MCCARTY EXPERIMENTS

Oswald Avery, the scientist who would later lead a groundbreaking experiment that would prove that DNA was genetic material, was born on October 21, 1877, in Halifax, Nova

Scotia. His family moved from Canada to the United States, where Avery and his two siblings grew up, in New York City on the Lower East Side. His childhood was cut short, however, by a series of tragedies; his oldest brother Ernest died at the age of eighteen and his father passed away a few months later. Avery himself was only fifteen years old and had to assume his father's role as the head of the family.

After graduating from Colgate University, Avery studied to be a doctor at the College of Physicians and Surgeons in New York. Inspired by his work with patients, Avery decided to study bacteria more closely at the Hoagland Laboratory in Brooklyn. There, he studied the bacteria in yogurt and tuberculosis, a bacterial infection of the lungs and one of the leading causes of death at the time. He developed a reputation for being able to break down the chemical composition of various bacteria.

By 1944, most scientists believed that genetic material was carried in the nucleus of the cell, but they didn't know which component of the nucleus—the proteins, DNA, RNA, or carbohydrates—carried the genetic information. Avery and his colleagues settled the matter once and for all with their groundbreaking experiment in 1944. They took two types of viruses, streptococcus pneumonia type IIIS and streptococcus pneumonia type IIR. They killed the streptococcus pneumonia type IIIS with heat and put those dead cells in the same dish as the type IIR virus. Lo and behold, within a few days, some of the type IIR virus cells had begun reproducing type IIIS virus cells. Somehow, the genetic information from the dead type IIIS cells had passed on to the type IIR cells. But how?

The scientists tested each of the components of the cell in turn by using **enzymes**. For example, they would take protease to eliminate the proteins in the cell nucleus of the type IIIS cells and put them in the dish with the type IIR cells. The type IIR cells still took in genetic material from the type IIIS cells, which meant that the genetic information wasn't in the protein. They eliminated each cell component, one by one, until they

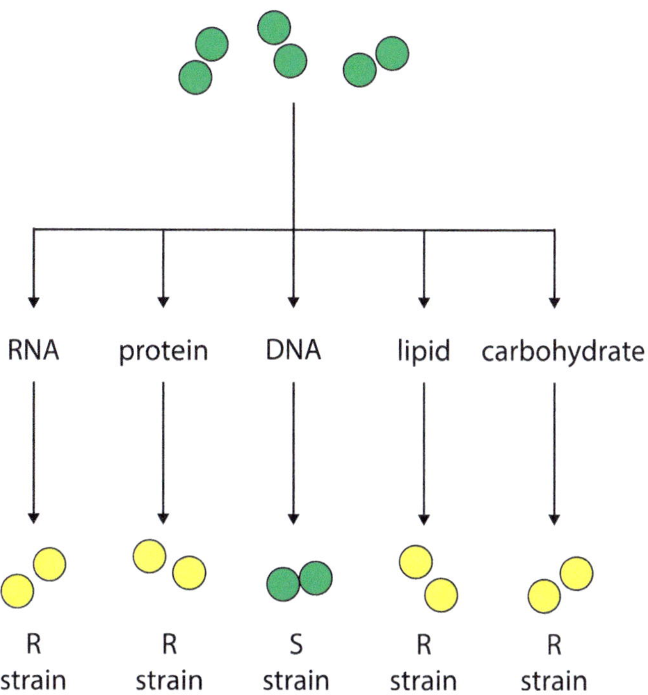

The Avery-MacLeod-McCarty experiment proved that DNA carries genetic material.

got to DNA. They used the enzyme DNAse to get rid of the DNA of the type IIIS cells. When those cells were transferred into a petri dish with the type IIR cells, they did not transfer their genetic material. Bingo! The experiment was the first to prove that DNA carried genetic material.

Several scientists seemed to confirm the results of the Avery-MacLeod-McCarty experiment, including French microbiologist Andre Boivin and American geneticist Edward Lawrie Tatum.

In 1952, the famous Hershey-Chase experiments conducted by Alfred Hershey and Martha Chase made it clear to scientists all over the world that it was indeed DNA that carried genetic material. At the time, scientists knew that viruses attached to bacteria cells and shot their genetic material into the cell. The genetic material from the viruses would

Martha Chase

Martha Chase was one-half of the duo that would confirm the Avery-MacLeod-McCarty experiment results with the Hershey-Chase study. Chase was born in Cleveland, Ohio, in 1927. Her father was a professor at a medical school. She had an early love for science and even conducted her own genetic experiments on fruit flies while she was getting her undergraduate degree at the University of Wooster. When she was only twenty-one years old, she joined a group of leading scientists at the legendary Cold Spring Harbor laboratory. The scientists formed a club called the **Phage** Group. Their goal was to learn more about genetics through bacteriophages, or viruses that infect bacteria by transferring genetic information to them. Chase helped design the famous Hershey-Chase experiment, which would elegantly prove through radioactive phosphorous and sulfur that it was DNA, not proteins, that carried genetic material.

Unfortunately, after the experiment, Chase suffered from many health and career difficulties. She was laid off from her research job and then went through a messy divorce from scientist Richard Epstein. She moved back to live with her father in Ohio. Many say that Chase's career was cut short because she was a woman. Three male scientists she had worked with in the Phage Group, Alfred Hershey, Salvador Luria, and Max Delbruck, would later go on to win Nobel Prizes, while Chase's own contributions have largely been forgotten.

combine with the genetic material in the cell to create infectious bacteria. Hershey and Chase found a way to add radioactive sulfur and phosphorous inside viruses. If you remember, DNA contains a lot of phosphorous, but no sulfur, and proteins contain a lot of sulfur, but no phosphorous. The researchers knew that if they found a lot of radioactive phosphorous in the bacteria, then it was DNA that carried genetic material. On the other hand, if they found a lot of radioactive sulfur in the bacteria, then it was proteins that carried genetic material. In their Nobel Prize–winning experiment, they found that the bacteria cells were full of phosphorous, proving once and for all that it was DNA that carried genetic information.

The STRUCTURE of DNA

Another big question about DNA was how the molecule was structured. The structure of any molecule tells you how it works. Knowing the structure of DNA would help scientists better understand and maybe even manipulate DNA. But figuring out the structure would take a long time. In the coming years, everyone from Russian-born Phoebus Levene to Viennese scientist Erwin Schrodinger would offer up their suggestions on what DNA might look like. Even though those early theories were incorrect, they would set the stage for the big discovery to come.

Phoebus Levene and Nucleotides

Phoebus Levene was born Fishel Aaronvich Levin in Sagor, Russia. He studied medicine at the Imperial military medical academy. In 1893, he moved his family to New York City, where he practiced medicine on the Lower East Side. Levene became interested in biochemistry and began studying the chemical structure of proteins at Columbia University, where he met Albrecht Kossel (who, if you remember discovered the

Phoebus Levene was a Russian-American biochemist who proposed an incorrect theory for the structure of DNA.

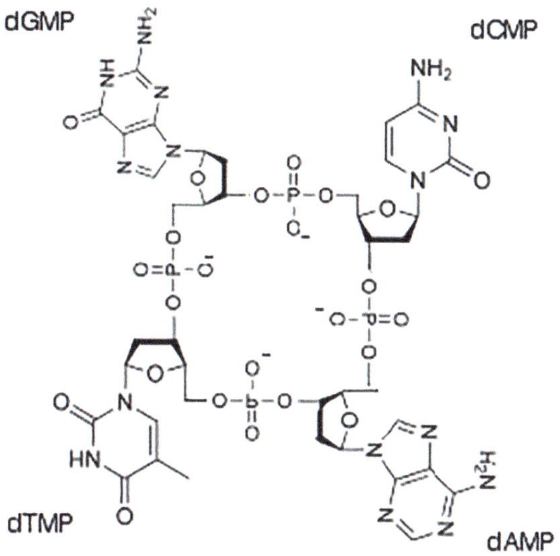

The tetranucleotide theory proposed that each DNA molecule had exactly one of each nucleobase hanging off the corner. It was later proven to be incorrect.

nucleobases in DNA and RNA). While at Columbia, Levene also published several papers on the structure of proteins. Because of his work, Levene became the head of the biochemistry lab at the Rockefeller Institute of Medical Research in 1905.

Through a series of experiments at the Rockefeller Institute, Levene discovered two sugars—ribose in 1909 and deoxyribose in 1929. These two sugars make up the backbones of DNA and RNA. He also discovered that DNA was formed by a series of phosphate-sugar-nucleobase chains, which he called **nucleotides**. Levene thought that each piece of DNA had only four bases, and he thought that it would be impossible for such a simple structure to hold genetic information. He would, of course, be proven wrong on both accounts.

Levene was the first to put forward a theory on the structure of DNA. His theory was called the tetranucleotide theory. Based on his experiment, Levene realized that there seemed to be equal amounts of adenine, thymine, guanine, and cytosine in each piece of DNA. This is why he suggested that DNA was shaped like a square, with one adenine nucleotide, one guanine nucleotide, one cytosine nucleotide, and one thymine nucleotide. Each of the nucleotides would be at one of the four corners of the square. Because Levene was a widely respected scientist at the time, many people believed in the tetranucleotide hypothesis for a long time. They also all believed that because DNA seemed to be simple, proteins were the real carriers for genetic information. Although this theory would eventually be proven incorrect, it set the stage for all the work that would be done on the structure of DNA in the very near future.

Erwin Schrodinger and the Irregular Crystal

The only child of Rudolf and Georgine Schrodinger, Erwin Schrodinger led a privileged childhood in Vienna. Both of his parents were interested in science—his father was a botanist and his mother's father was a professor of chemical technology at the Imperial Royal Technical University of Vienna. Schrodinger himself studied under private tutors at home. He studied at the Imperial Royal Academic High School, where he was a star student and graduated with honors in 1906.

For college, Schrodinger studied physics at the University of Vienna. There, he was inspired by Fritz Hasenohrl, a brilliant physicist who studied the relationships between an object's mass and its energy. When World War I broke out, Schrodinger was required to join the army because of Austria's mandatory military service requirement. While stationed in Prosecco, Vienna, he read Albert Einstein's theories, which were just being published and popularized. In 1920, Schrodinger was awarded the Haitinger Prize of the Austrian Academy of Sciences for his theories on color. The prize was awarded to the person whose "studies in chemistry and physics…proved to be of great practical use for industrial applications."

After the war, Schrodinger moved to Oxford University, then Princeton University, and continued to be a mover and shaker in the field of quantum theory—a field that uses physics to explain the movement of **atoms**.

In 1944, Schrodinger wrote his book, *What Is Life?*, which attempted to use physics to predict the structure of DNA. In the book, Schrodinger argues that chromosomes probably contain a full code that gives information for the characteristics of all living things. Schrodinger believed that genetics were in a very large protein molecule in the center of every atom. He also talked about previous work done on mutations in genes.

Schrodinger offered a possible structure for genes—he described them as an aperiodic crystal, which means he believed they had the solid structure of a crystal but didn't repeat themselves. This was different from the previous tetranucleotide theory structure that Levene had suggested. Schrodinger said the structure was probably non-repeating because it needed to be complex enough to include genetic information.

Linus Pauling, one of the leading biochemists of the time, said that Schrodinger's book was one of the most significant books in scientific history. The book was the first time anyone had applied quantum mechanics to issues in chemistry. It marked the beginning of molecular biology as a scientific field.

Schrodinger's book received mixed reviews because of his epilogue, where he extended his scientific theories to philosophical theories. He wrote that because the body is a force of nature that follows the laws of physics, and because we are all directing the movement of those atoms in our bodies, we are basically a force of nature. He argued this gave us a god-like quality. Many scientists disagreed with him and felt like he was overstepping his role as a scientist by making such bold claims about faith.

Erwin Chargaff and Chargaff's Rules

Erwin Chargaff was born on August 11, 1905, in Chernivtsi, in present-day Ukraine. His father, Hermann Chargaff, was a banker, and his mother was Rosa Silberstein Chargaff. During World War I,

Austrian biochemist Erwin Chargaff resigned from his post at the University of Berlin when the Nazis began excluding Jews from the university during World War II.

his father lost his bank, and later in World War II, his mother was targeted for being a Jew. Chargaff studied chemistry in Vienna and eventually earned his doctorate at Vienna University of Technology while studying under Fritz Feigl, who created the "spot test," a technique for analyzing chemicals in a simple way.

Chargaff eventually immigrated to the United States and became a professor at Columbia University, where he worked in the biochemistry department.

Chargaff became interested in DNA as genetic material after Oswald Avery's experiments in 1944. He started working with his graduate student, Ernst Vischer, on ways to separate nucleobases from the rest of the DNA molecule. They came up with a technique to separate material and color it, called **chromatography**. They found that in each strand of DNA, there were equal amounts of adenine as there were thymine, and equal amounts of cytosine as guanine. This became the basis of what was called Chargaff's rules. He announced those rules in a talk he gave in 1949.

The findings also helped Chargaff disprove Levene's tetranucleotide theory. Through chromatography and **spectrophotometry**, which helped them count the number of bases in each DNA sample, they were able to prove that there were actually different amounts of the nucleobases, even though there seemed to be a ratio between adenine and thymine and between guanine and cytosine.

JEAN BRACHET, TORBJÖRN CASPERSSON, and the RNA TIE CLUB

While there has been very little written about the history of RNA before the discovery of the DNA double helix, most historians agree that the discovery of RNA goes back to Miescher (the sticky substance he had found contained both DNA and RNA, but he didn't know it at the time). Over the next hundred years, most of the attention centered around DNA, but there were some scientists who were also studying RNA, a molecule very similar to DNA. RNA also has nucleobases, but instead of thymine, it is composed of a nucleobase called uracil. In 1933, Belgian biochemist Jean Brachet discovered that RNA, unlike DNA, is found in the cytoplasm of the cell. The cytoplasm is the entire thick, watery area outside of the cell's nucleus. Brachet was also able to show that chromosomes were made up of DNA.

Swedish scientist Torbjörn Caspersson also added to knowledge about RNA. Caspersson was born in Motala, Sweden. He studied medicine and eventually received his doctorate in 1936. Between 1937 and 1939, he conducted experiments that found that large amounts of RNA were found in cells as made proteins, which suggested that RNA was a critical component to the creation of proteins. This finding would become important later on, as a group of scientists who formed the RNA Tie Club would use the information to figure out the exact role RNA plays in the cell.

Caspersson also helped set the groundwork for discovering the structure of DNA. He worked in the lab of Einar Hammarsten, a chemist who was studying nucleotides (the building blocks of DNA). By using different filters, they found that nucleotide molecules were actually larger than protein molecules (which was contrary to past beliefs at the time). They were also able to figure out that DNA was a large molecule with thousands of nucleotides. This disproved Levene's tetranucleotide theory and other theories that suggested that DNA was very short. Scientists now know that DNA, that tiny stuff inside our cells, stretches out to 2 meters (or over 6 feet) long!

TAKING PHOTOS of DNA: The ADVENT of X-RAY CRYSTALLOGRAPHY

One of the single greatest breakthroughs in technology that would pave the way for the discovery of the structure of DNA was **X-ray crystallography**. It was invented by William

Lawrence Bragg and his father, William Henry Bragg. The new technology allowed scientists to take pictures of tiny molecules that they were not able to see before.

William Lawrence Bragg

William Lawrence Bragg was born in Australia in 1890 to William Henry Bragg and Gwendoline Bragg. His father was an accomplished physicist, and the father/son duo would eventually share a Nobel Prize together. When he was five years old, the younger Bragg fell off his bike and broke his arm. His father, who had read about the newly emerging X-ray technology, took an X-ray of his son's broken arm. It is reportedly the first X-ray taken in Australia.

The younger Bragg was extremely bright and graduated from high school when he was only fourteen years old. He graduated from college at eighteen. In 1909, his father accepted a position as the chair of the physics department at the University of Leeds. The younger Bragg received a scholarship to study mathematics at Cambridge and eventually graduated with a degree in physics.

Around 1912, Bragg and his father learned about the work that German physicist Max Theodor Felix von Laue was doing on how crystals bounced X-rays off their structure. After World War I, they started to develop a technique called X-ray crystallography, the process of shooting X-rays at crystal structures to make patterns with rays. The method was based off of a theory developed by the younger Bragg called Bragg's law. Bragg's law is a mathematical formula that helps scientists figure out the spacing between the layers of crystals. The law helps scientists figure out the location of very small particles based off of X-rays. Imagine throwing a tennis ball at something invisible. If you threw enough tennis balls, you could figure out where this invisible object was and how big it was, even if you couldn't see it because you could observe how the tennis balls bounced off of the object and infer its location. Because we cannot physically see molecules with the naked eye, X-ray crystallography helps scientists figure out what very small structures look like. A lot of salts, minerals, proteins, and metals form a crystal structure, so this technique was a huge step for science.

Linus Pauling, James Watson, and Francis Crick would all later use X-ray photographs from X-ray crystallography to build physical models of DNA as they tried to figure out its structure. They used the photographs to figure out the spacing and the bond angles of the different atoms.

William Astbury and the X-Rays

While Bragg discovered the X-ray crystallography method, it was William Astbury who would take the first steps to using the method on DNA. Astbury came from a poor family—his father supported four children as a potter. Yet Astbury managed to get a superior education by winning scholarships for high school and college, an early sign that he was destined for great things. In 1917, he received a scholarship to Jesus College in Cambridge to study physics and mathematics, but his studies were interrupted by the First World War. He served in Ireland and finished his studies after the war ended. Astbury took a class in crystallography, the study of the structure of complex molecules. The professor was so impressed by Astbury that he recommended the student to Sir William Lawrence Bragg, the inventor of the X-ray crystallography technique.

Astbury would eventually surpass Bragg's own technique in X-ray crystallography. He was such a great X-ray crystallographer that Nobel Prize–winning molecular biologist Max Perutz once dubbed his lab "the X-ray Vatican." In the beginning, Astbury took lots of photos of an element called tartaric acid, an organic acid that occurs in a lot of plants, including grapes. Although he wasn't able to figure out the

William Lawrence Bragg found out that he had received the Nobel Prize in Physics around the same time he found out that his brother had been killed at war.

structure of the acid, his early papers showed how he drew metaphors between unknown structures (like tartaric acid) and known structures (like ice) to fill in the blanks. This mode of thinking would serve him well in the future.

Next, the textile industry paid Astbury to study the proteins in wool fibers, including keratin and collagen. By the 1930s, Astbury was able to show that wool stretched by almost twice its size when wet. This also affected the coil structure of the proteins that make up wool. Astbury proposed that there were two different forms of structure for proteins—stretched, which he called β-form, and unstretched, which he called α-form. His photographs also helped him figure out that proteins formed spiral coils and folded over themselves. All these findings would later be critical to American scientist Linus Pauling, who ultimately figured out the structure of proteins.

All this work led up to Astbury's greatest contribution to the discovery of the structure of DNA. Astbury had suspected for a long time that DNA was very important. As he said in his Harvey lecture in 1950:

> The proteins lie at the corner of the business, we may be sure of that, and the attack on them must go on unceasingly; but they are not, or have not come to be, entirely self-acting, and it is little less urgent—it is a parallel problem—to concentrate also on their collaborative macromolecules, notably the polysaccharides and **nucleic acids**, especially the nucleic acids. Many investigations in recent years have brought out and emphasized the importance of the nucleic acids in biosynthesis; as far as we can see, they are absolutely essential components in chromosomal processes and cell multiplication and in virus reproduction, and in fact it is largely believed now that the nature of the interaction of the proteins and

nucleic acids is probably the supreme issue of all in the chemistry and physics of life.

In 1937, the Swedish scientist Torbjörn Caspersson sent him DNA from calf thymus to study. It was then that Astbury took one of the first X-ray photographs of DNA. Unfortunately, the technology was still not well developed yet and Astbury couldn't make out the helix structure of DNA—it looked like the molecule was arranged in crosses. However, his photographs and his work would eventually be the basis for the photographs that later allowed James Watson and Francis Crick to figure out the structure of DNA.

The EUGENICS MOVEMENT

Along with the growing knowledge and interest in genetics, the United States gave birth to the eugenics movement in the late nineteenth and early twentieth centuries. Eugenics is the practice of cultivating the gene pool by preventing certain people from having children and thus preventing them from passing on their genes. Many of these traits that eugenicists wanted to get rid of were vague, like "feeble-mindedness," or simply those deemed "defective." They believed that they could make the human race stronger by breeding out these traits. Many leaders of the eugenics movement used science to justify racist policies—they believed that white, Nordic people carried the superior genes and everyone else should be prevented from having children. Surprisingly, however, there were also several black American activists who agreed with them—although they believed that the "best" black Americans had genes that were just as valuable as white genes and proposed that the "best" black people be bred with the "best" white people.

The American eugenics movement received a ton of support from famous donors like the Rockefeller Foundation, the Carnegie Foundation, and others. In 1906, Charles

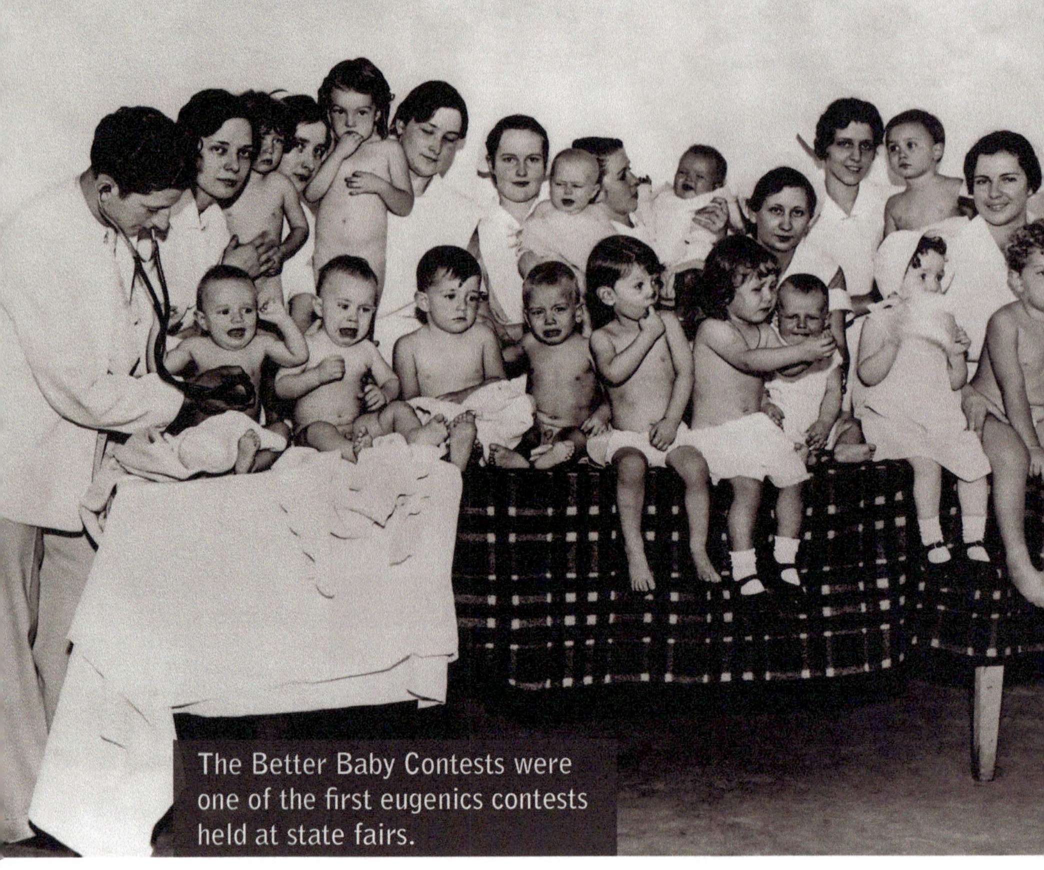

The Better Baby Contests were one of the first eugenics contests held at state fairs.

Davenport, one of the most outspoken leaders of the eugenics movement, established the American Breeder's Association, whose mission was to "investigate and report on heredity in the human race, and emphasize the value of superior blood and the menace to society of inferior blood." In 1896, Connecticut enacted one of the first eugenics laws, which prohibited anyone who was "epileptic, imbecile, or feeble-minded" from marrying. In 1907, Indiana enacted the first "sterilization legislation." Between 1907 and 1947, Indiana performed 2,500 sterilizations. In 1927, the Supreme Court case *Buck v. Bell* legitimized the forced sterilization of patients at a Virginia home for the intellectually disabled. In North Carolina, social workers were able to approve sterilization for anyone with an IQ of 70 or lower.

In 1908, Mary DeGarmo, a former classroom teacher, created the "Better Baby Contest." DeGarmo partnered with a physician, Dr. Jacob Bodenheimer, to find the baby with the "perfect genes." Babies were judged for their physical and intellectual development. These contests, which were held at state fairs, became extremely popular. Sometimes they were even the most popular event at the fair! The NAACP even held Better Baby contests as fund-raisers. Later on, state fairs held Fitter Family Contests, where families were judged based on their size, overall attractiveness, overall health, and overall intelligence.

The eugenics movement may have continued if it wasn't for World War II. Startled by Adolf Hitler's extreme views on the superior human race, the study of eugenics quickly fell out of favor in the United States. However, increased knowledge of genes and the manipulation of genes would continue to pose ethical questions about how far this science should be used to alter the human population.

DNA IS the GENETIC MATERIAL, NOW WHAT?

In less than a few decades, interest in DNA began ramping up quickly. A series of experiments proved definitively that it was DNA, not proteins, that held genetic information. Technology also advanced so that scientists could take pictures of tiny structures inside the body they were not able to see before. While they still didn't know how DNA was structured, or how exactly it did its job, they were closer than ever to cracking the code. It was only a matter of time.

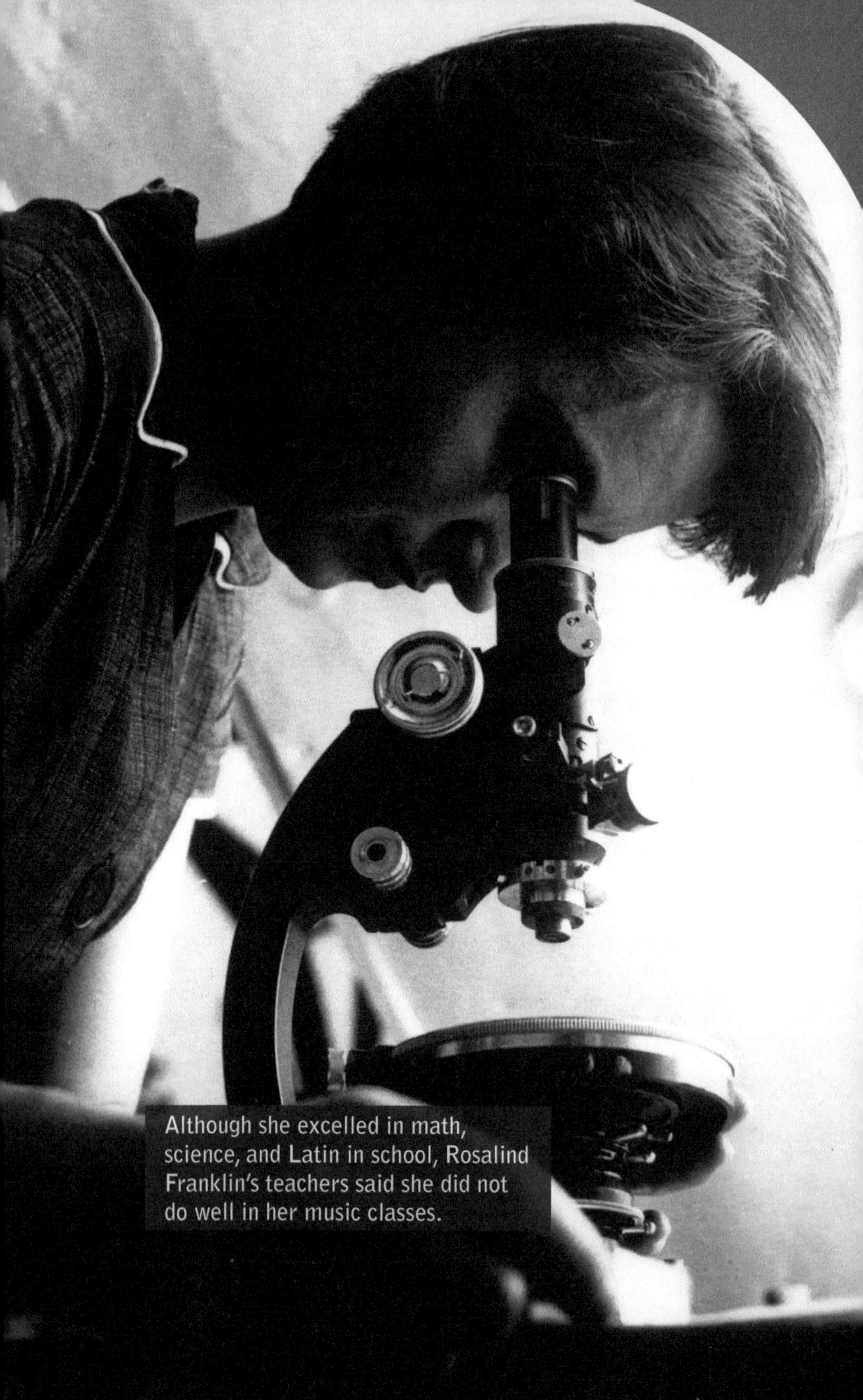

Although she excelled in math, science, and Latin in school, Rosalind Franklin's teachers said she did not do well in her music classes.

CHAPTER 3

The Major Players in the Discoveries

Discovering the structure of DNA took the combined efforts of four very talented scientists. Despite a friendly rivalry, the scientists worked closely together, and without each of their individual efforts it may have taken a lot longer for the world to know the structure of DNA. Those four scientists, James Watson, Maurice Wilkins, Rosalind Franklin, and Francis Crick, all came from different backgrounds, but they had one thing in common: a strong desire to discover the secret to life.

JAMES WATSON

April 6, 1928, was an unusually rainy day for Chicago. The newspapers worried that the weather would be bad for the wheat crops. It was on this day that James D. Watson was born to Margaret Jean Mitchell and James D. Watson Sr.

As a child, Watson was a well-behaved son who didn't like to comb his hair. He grew up on the South Side of Chicago, a working-class district full of factories, steel mills, and meat-packing plants. Watson devoured the *World Almanac* for fun and once won $100 on a popular radio-show trivia program called "Quiz Kids." The radio show brought in panels of five

James Watson loved learning from an early age and started college at the age of fifteen.

children chosen for their intelligence and personalities to answer questions sent in by listeners.

On Sunday mornings, Watson joined his father and his father's friends on bird-watching expeditions instead of going to church, until he was twenty-two years old. "Birds dominated my father's life," Watson said. Although his family had a long history of Catholicism (his mother's family were Irish and Scottish Catholics), Watson himself would later renounce religion and remark that he was "extremely lucky" that his father didn't believe in religion. Watson's own interest in birds grew, and he dreamed of studying them as an ornithologist when he grew up. Watson also learned of Charles Darwin as he growing up, and he has said that Darwin's theories of evolution made a huge impression on him. "Early on, I heard of Charles Darwin," he said in his TED Talk in 2007. "I guess, you know, he was the big hero. And…you understand life as it now exists through evolution."

When young Watson was fifteen years old, a man named Robert Hutchins had just become the president of the University of Chicago. Hutchins was an ambitious educator, known for his big ideas and tendency to rub people the wrong way with his air of arrogance. When he arrived at the University of Chicago, he immediately began putting his controversial ideas in motion. Among them was the idea of starting kids with a college education at the age of fifteen. Hutchins believed that the purpose of education was not simply to prepare students to earn a living and do a job, but to teach young people how to think. He also believed that kids were at their most impressionable in their teenage years and should be taught skills of critical thinking early on. This idea became reality in the form of the Gifted Youngster Program. Watson applied to the program and received a tuition scholarship to study zoology.

During his last year of college however, Watson's interests changed completely. He had picked up the book, *What Is Life?* by the Austrian physicist Erwin Schrodinger, and it opened the world of genetics to him. The book was based off of a series of lectures Schrodinger had given at the Dublin Institute for Advanced Studies at Trinity College, a liberal arts college in Dublin, Ireland. In his book, Schrodinger discussed the fact that whatever carried the **hereditary** information must be small enough and replicable across time, which went against the principles of classical physics (which stated that the smaller atoms become, the more random and unpredictable they become). Schrodinger hypothesized that whatever carried genetic information was probably a molecule rather than an atom. Molecules were small enough and relatively stable compared to atoms. Schrodinger predicted that molecules would have the stability of a crystal-like structure. With this book, Schrodinger was able to set the theoretical groundwork for the discovery of the structure of DNA. Schrodinger's writing and theories were so captivating to the eighteen-year-old Watson that he switched his professional focus from ornithology to genetics.

At the time, the famous scientist Linus Pauling was one of Watson's biggest heroes. He applied to work at Caltech, where Pauling was conducting research, but was turned down. This is why Watson applied to do his Ph.D. at Indiana University after graduating from the University of Chicago. Indiana University was the home of Nobel Prize–winning geneticist Hermann Joseph Miller. There, Watson found a mentor in Salvador Luria, an Italian-born geneticist who studied the genetic structure of viruses. Under Luria's guidance, Watson wrote his Ph.D. thesis on how hard X-rays—the highest energy X-rays—could be used to observe bacteriophage multiplication. Bacteriophages are viruses that infect bacteria and pass on their genes. When the bacteria multiply, they replicate the genes they are carrying. Many years later, bacteriophages would also be critical to the process of artificially replicating genes.

Afterward, Watson spent a year at Copenhagen University in Herman Kalckar's lab. Unfortunately, Kalckar wasn't interested in exploring the structure of DNA, which was Watson's main interest, so Watson moved over to work with microbial physiologist Ole Maaloe, then a member of the Phage Group. The Phage Group (as discussed in chapter 2) was an informal club of biologists including Watson's old mentor Luria, as well as Alfred Hershey, Seymour Benzer, Gunther Stent, Frank Stahl, and Renato Dulbecco, all of whom studied bacteriophages.

In the summer of 1951, when Watson accompanied Kalckar to a meeting in Italy, he met a young, square-jawed New Zealand-born British physicist named Maurice Wilkins. Watson saw the beautiful X-ray diffraction photographs Wilkins had taken that revealed the crystalline structure of DNA. With that, Watson became more determined than ever to discover the structure of DNA. That fall, Luria arranged for Watson to begin working at the Cavendish laboratory, a physics lab at the University of Cambridge. Watson had no idea how soon he and Wilkins would be working together.

Maurice Wilkins shared a Nobel Prize with James Watson and Francis Crick in 1962 for the discovery of DNA.

MAURICE WILKINS

In many ways, Pongaroa, New Zealand, was the edge of the world. It was a tiny, rural, sheep-filled countryside, hundreds of miles away from any large cities or towns. It was home to aboriginal tribes, including the fierce Ngāti Kahungunu, one of the largest tribes in New Zealand, and a few white settlers. On December 15, 1916, Maurice Hugh Frederick Wilkins was born in the middle of this countryside. Wilkins often said that he felt the pioneering spirit of the New Zealand countryside served him well later in life.

When Wilkins was six years old, his father, who was a medical doctor, decided that the New Zealand countryside wasn't the ideal place for him to launch the career he wanted. He moved the family to Birmingham, England, to pursue a career in preventive medicine. In 1935, Wilkins began studying physics at St. John's College at Cambridge and went on to get a Ph.D. at Birmingham University while studying phosphorescence, which is a light that is emitted when particles absorb radiation.

When Wilkins left school, World War II had just broken out. England was being attacked by Germany, and Wilkins was able to put his research to use immediately. He used his research on phosphorescence to improve the brightness of radar screens. This let the Allied armies spot enemy airplanes, ships, and submarines hundreds of miles away, even at night. He also worked with Professor M.L.E. Oliphant, an Australian physicist and one of the first scientists to come up with the idea of making atomic bombs by separating different forms of uranium atoms. Wilkins, Oliphant, and a group of scientists who were doing this research together eventually moved from Birmingham to the Manhattan Project in Berkeley, California.

The scientists involved in the Manhattan Project thought they were helping to bring about peace in the world by deterring the Nazis from continuing the war. Imagine their shock when they learned that their research was instead being used to kill innocent Japanese civilians. After the war, Wilkins vowed to dedicate his life to the science of life instead of the science of destruction. In 1963, he told the *Sunday Review* magazine, "Partly on account of the [atomic bomb], I lost some interest in physics." Later, Wilkins would be remembered for his gentle nature and his outspokenness toward war. "He was a very intelligent scientist with a very deep personal concern that science be used to benefit society," Watson said after Wilkins passed away. "This started in his early days, when he witnessed the atrocities of war, and continued through his life. He will be sorely missed."

After the war, Wilkins looked for other fields of science to turn to. Like Watson, he became enthralled by Schrodinger's book *What Is Life?* When scientists at the Rockefeller Institute in New York determined that DNA was potentially the source for genetics, the physicist jumped at the chance to study it. He prepared to turn his prior training from physics to the field of biology.

A man by the name of John T. Randall had just started a biophysics unit at King's College in London in 1946, and he

Maurice Wilkins was one of the scientists who worked on the atomic bomb during World War II. After the war, Wilkins turned his attention to the study of DNA.

asked Wilkins to join him. The two had worked together before at the University of St. Andrews (Randall had been the chair of the department, and Wilkins had been his assistant lecturer).

Wilkins agreed and moved to King's College, London, where he began experimenting with ways to X-ray DNA. A Swiss scientist by the name of Rudolf Signor had given him some strands of DNA from a calf thymus. Wilkins noticed that when he put a rod inside the jar of DNA, the thin fiber would wind itself around the rod. Wilkins realized that the spiral nature of the fiber would be perfect for X-ray crystallography and proceeded to give it a go.

At the time, Wilkins was friends with another scientist by the name of Francis Crick. Crick was dismissive of Wilkins's initial interest in DNA. He told Wilkins to "find yourself a good protein" to study instead.

Despite his naysayers, Wilkins continued to study DNA. His first breakthrough occurred when he and his assistant Raymond Gosling separated DNA into small fibers, mounted those fibers on a paperclip, and fired X-rays at them. The X-ray images showed that what was previously thought of as simply a gooey, sticky substance actually had a crystalline structure at the atomic level. The method of X-ray that Wilkins used was a more refined version of X-ray crystallography, the method that Sir William Lawrence Bragg has discovered in 1912. "When ... I first saw all those discrete diffraction spots ... emerging on the film in the developing dish was a truly eureka moment...we realised that if DNA was the gene material then we had just shown that genes could crystallize!" Gosling later said. Wilkins believed that if he got enough photos of the crystal pattern of DNA, he could figure out the structure of DNA itself.

In 1951, the same year Wilkins met Watson at the conference in Italy, he would also meet a feisty young woman who would soon throw some complications into his research. That woman's name was Rosalind Franklin.

ROSALIND FRANKLIN

Rosalind Elsie Franklin was born on July 25, 1920, in Notting Hill, London, England. Her family was wealthy and influential. Her father's family had started an investment bank in Britain called Keyser and Company. Her father worked at the bank and taught at the Working Men's College, one of the oldest adult-education centers in Britain. Franklin's aunt and uncle held prominent seats in the British government.

Franklin was the second out of five children. She was very intelligent as a child: her aunt reported that at six years old, Franklin would spend all her time doing math for fun. She got all the problems right. When Franklin was eleven years old, her parents enrolled her in Saint Paul's Girls School, one of the leading day schools for girls in Britain. She was always the top student in all her classes (except music, where her teacher said she was tone deaf). Inspired by her classes, Franklin decided she wanted to be a scientist when she was fifteen years old.

As much as Franklin wanted to be a scientist, however, her father thought it was a bad idea. At the time, it was very difficult for women to become scientists, and he worried she wouldn't be able to find a job. Society still believed that a woman's place was in the home and that a woman wouldn't be able to balance a successful scientific career with family.

Despite her father's concerns, Franklin enrolled at Newnham College, one of the two women's colleges at Cambridge University at the time. She studied chemistry and continued to excel in all her classes. In 1941, the college awarded her second-class honors on her final exams, which were equivalent to a bachelor's degree for women at the time. After leaving Cambridge, Franklin received a research fellowship to work with Ronald Norrish, a scientist who would later go on to win the 1967 Nobel Prize in Chemistry. Unfortunately, Norrish and Franklin didn't get along. Franklin was used to working independently and didn't like taking orders from Norrish. For his part, Norrish didn't think that women

Letter from John T. Randall to Rosalind Franklin, asking her to join his lab and start studying DNA:

4th December, 1950

Dear Dr. Franklin,

I am sorry I have taken so long to reply to your letter of November 24th. The real difficulty has been that the X-ray work here is in a somewhat fluid state and the slant on the research has changed rather since you were last here.

After very careful consideration and discussion with the senior people concerned, it now seems that it would be a good deal more important for you to investigate the structure of certain biological fibres in which we are interested, both by low and high angle diffraction, rather than to continue with the original project of work on solutions as the major one.

Dr. Stokes, as I have long inferred, really wishes to concern himself almost entirely with theoretical problems in the future and these will not necessarily be confined to X-ray optics. It will probably involve microscopy in general. This means that as far as the experimental X-ray effort is concerned there will be at the moment only yourself and Gosling, together with the temporary assistance of a graduate from Syracuse, Mrs. Heller. Gosling, working in conjunction with Wilkins, has already found that fibres of desoxyribose nucleic acid derived from material provided by Professor Signer of Bern gives remarkably good fibre diagrams. The fibres are strongly negatively birefringent and become positive on stretching, and are reversible in a moist atmosphere. As you no doubt know, nucleic acid is an extremely important constituent of cells and it seems to us that it would be very valuable if this could be followed up in detail. If you are agreeable to this change of plan it would seem that there is no necessity immediately to design a camera for work on solutions. The camera will, however, be extremely valuable in searching for large spacings from such fibres.

I hope you will understand that I am not in this way suggesting that we should give up all thought of work on solutions, but we do feel that the work on fibres would be more immediately profitable and, perhaps, fundamental.

I think I must leave to you the question as to whether you come over here for a day or two to discuss these matters further. It now seems so near to the time when you will actually be working here that it is perhaps hardly necessary for you to make the special journey. On the other hand, there may be things which you could organize on the apparatus side in Paris and you could hardly do this without further discussion with us. The change of program, such as I have suggested, will probably mean that we should obtain the formal consent of the Fellowship Committee; there is no hurry about this and there is no doubt about the answer.

Dr. Price has just heard from Mr. Heins of the Rockefeller Foundation that orders have now been placed for your apparatus.

Yours sincerely,
J.T. Randall

scientists should have equal say in the lab. Eventually, unable to handle it anymore, Franklin resigned from Norrish's lab.

Franklin next applied for a job at the British Coal Utilization Research Association in 1942. The research group was started to explore the use of coal and charcoal as fuel for wartime vehicles. Charcoal acts as a filter for toxins when activated with oxygen so scientists were looking for ways to use it in gas masks. Franklin contributed a large amount of knowledge to this field of research. By studying a variety of different types of coal in the British Isles, Franklin was able to figure out which types of coal had the largest and smallest holes (the size and shape of the holes in coal affect how quickly it will burn). Her work allowed the army to group coal by how easy it would be to burn. She received her Ph.D. in 1945 for her research on coal.

After the war, Franklin joined the Laboratoire Central des Services Chimiques de l'Etat in Paris, a lab funded by the French government, where she worked under Dr. Jacques Mering. Mering taught Franklin the art of X-ray crystallography, a relatively new technique that allowed scientists to photograph tiny crystals by sending X-ray beams through them. The X-rays produced a pattern of spots on a photographic plate, which scientists then ran through several mathematical equations to determine the structure of the crystal. Franklin became an expert in the complex and time-consuming technique of X-ray crystallography, and many scientists have said that she produced some of the most beautiful X-ray diffraction photographs ever done. She used the technique to further explore coal and carbon and was able to publish more papers on the unique characteristics of each substance, which she discovered through X-ray.

While Franklin enjoyed the work, she didn't want to photograph pieces of coal forever. Her friend Charles Coulson, a theoretical physicist, suggested that she should use her expertise to study larger biological structures like DNA. Coulson knew that DNA was starting to get a lot of attention, and if Franklin put herself in the right place at the right time, she could be part of something big. Taking her friend's advice, Franklin applied for and received a three-year Turner and Newall Fellowship to work as a research associate at King's College in London. The director of the biophysics unit, John Randall, wanted to use her experience in X-ray diffraction technology to study DNA. Randall paired Franklin with graduate student Raymond Gosling, who was then working with Maurice Wilkins.

Rosalind Franklin was a brilliant woman, one of the few women scientists of her day. She made a huge contribution to the discovery of DNA's structure, but unlike her male colleagues, she did not receive a Nobel Prize for her work.

FRANCIS CRICK

Francis Crick was born in Weston Favell, a small village near the English town of Northampton. His father managed a shoe and boot factory. His mother was a schoolteacher. Crick was very interested in science as a young boy. He read the entire series of the *Children's Encyclopedia*. Based on what he read, he tried to conduct his own science experiments in his kitchen (he tried to make artificial silk but failed). When he was fourteen years old, Crick received a scholarship to the Mill Hill School, one of London's leading independent boarding schools, where he studied mathematics, physics, and chemistry. It was here that Crick immersed himself in learning and first got interested in the science of inheritance. He read about the genetic experiments that Gregor Mendel had conducted many years previously on pea plants and taught himself the Mendelian laws of genetics. Crick often credited his later success to the education he received at the Mill Hill School.

He later said that he had an early interest in science in an interview with Web of Stories. "As far I remember I was interested in science. I didn't ask myself what happens when

After the discovery of DNA, Francis Crick wrote a book called *Of Molecules and Men*, where he talked about the relationship between religion and science.

ice melts for example," he said. "But when I read about it later I thought, isn't it interesting that we can think about things in nature in that particular way?"

Crick went to the University College in London and earned a bachelor's degree in physics in 1937. He started studying the viscosity of water—the science of how water molecules attract each other and hold together—for his Ph.D. but when World War II broke out, it was hard to continue his research. One morning during the famed Battle of Britain, a bomb crashed through the roof of his laboratory and destroyed his lab equipment. Eventually, Crick joined the Admiralty, the organization in charge of the British navy, where he helped design magnetic and acoustic mines. These were mines placed underwater to destroy ships and submarines and were very important in helping the British win the war.

None of this was truly exciting to Crick, however. In order to decide his next move, he asked himself what he liked talking to his friends about, and he settled on two topics: the biological basis of life and psychology. Crick eventually decided to study the biological basis of life (or DNA) because he felt that would be a better use of his previous experience.

Crick spent the next few years making up for lost time in the field of biology. He got some money to study at Strangeways Laboratory at Cambridge University, which was led by Nobel Prize winner Max Perutz. He eventually wrote his Ph.D. thesis on X-ray diffraction of proteins. In 1951, Crick was invited to join the Cavendish Laboratory headed by William Lawrence Bragg. There, he met a twenty-three-year-old young man named James D. Watson.

LINUS PAULING

Linus Pauling was born on February 28, 1901, in Portland, Oregon, to a poor family. His father started out as a bread salesman but after Pauling's sister was born, his father became a salesman with the Skidmore Drug Company. Pauling's father passed away when the boy was just nine years old, and the family, who was already struggling with finances, skidded to the brink of poverty.

A friend of Pauling's had a toy chemistry set, and Pauling became fascinated with the science. He went home and created a chemistry lab of his own in his basement. When he was sixteen years old, he dropped out of high school to study chemical engineering at Oregon Agricultural College. He knew so much about chemistry that he was sometimes asked to teach classes when there wasn't a teacher available. He met Ava Helen, who would later become his wife, when he was teaching one of those classes.

After graduating from Oregon Agricultural College, Pauling moved on to take a professorship at California Institute of Technology (Caltech). There, he published more than fifty papers and made huge contributions to the study of chemistry. Pauling used the new technology of X-ray diffraction to figure out the distances of atoms and came up with Pauling's rules, which were four rules about chemical bonds. One of his most important books was called *The Nature of the Chemical Bond*, which he also wrote at Caltech.

Pauling was initially interested only in studying inorganic molecules—all substances that don't contain carbon. However, he switched to studying organic compounds, especially proteins, at the suggestion of Warren Weaver, who was the head of the Rockefeller Foundation at the time.

Pauling was the youngest person to be elected into the National Academy of Sciences. He was one of the only scientists to win the Nobel Prize in two different categories: chemistry and peace. He was well respected as one of the smartest scientists of all time. James Watson, Francis Crick, Maurice Wilkins, and Rosalind Franklin all knew of his work. Francis Crick considered Pauling to be one of the founders of molecular biology, the science that includes the study of genetics.

William Lawrence Bragg refined the X-ray technique that was crucial for the discovery of the structure of DNA.

CHAPTER 4

The Discovery of DNA and RNA

By the mid-twentieth century, scientists knew that DNA existed, that it coded for genes, and that it was made up of sugars, phosphates, and nucleobases. But there was still a huge problem: no one knew how DNA was able to replicate itself because they didn't know the actual way that DNA was structured. How DNA actually coded the instructions for creating new cells in living organisms was also still a mystery.

Over the course of three years, world-class scientists, including Linus Pauling, Erwin Chargaff, Rosalind Franklin, and Maurice Wilkins, raced to figure out the structure of DNA. Many of them came close, but it was an unlikely pair who finally cracked the code. On February 28, 1953, more than one hundred years after Gregor Mendel's experiments in the 1800s, the search for the secret to how DNA coded genes finally came to an end. James Watson and Francis Crick, two scientists at the research group in Cavendish Laboratory at the University of Cambridge, published a paper with the solution to the mysterious substance of deoxyribonucleic acid. The paper was called "The Molecular Structure of Nucleic Acids." It was about to change the world.

Francis Crick and James Watson met in 1951 when Watson was visiting Cambridge. The pair hit it off immediately.

The STAGE is SET

As you might remember, Swiss biologist Friedrich Miescher first discovered DNA, which he called nuclein. He found a substance that was different from proteins and contained a large amount of phosphorous. But it wasn't until Oswald Avery and his colleagues conducted their experiments in

1944, that scientists suspected that DNA contained genes that coded for life. Even up until the late 1940s scientists were skeptical of Avery's work. Linus Pauling, Nobel Prize–winning American biochemist and the "father of biology," believed that DNA played only a minor role in coding life. However, by the 1950s scientists started to come to a consensus that DNA was probably the molecule most important for genetic information.

Phoebus Levene was one of the first scientists to make headway into how DNA is structured. He broke down yeast DNA and discovered that DNA was made up of chains of molecules called nucleotides, which were made up of one of the nucleobases (adenine, thymine, guanine, cytosine), a sugar, and a phosphate group. He even put forth one of the first hypotheses on how DNA was structured with his tetranucleotide hypothesis. He believed that all DNA was structured in the same order—guanine, cytosine, thymine, adenine—and that this order repeated itself. However, this was later proven wrong by Erwin Chargaff, the Austro-Hungarian biochemist, who proved through his experiments that the order of DNA bases is different for different species. He also proved that there was the same amount of adenine as thymine and the same amount of cytosine as guanine within DNA. These characteristics of DNA are known as Chargaff's rules, and they would be pivotal in the ultimate discovery of the structure of DNA.

Next, British physicist William Astbury took X-ray crystallography to the next level. He first took photos of proteins. After Swedish scientist Torbjörn Caspersson sent him DNA from a calf thymus, Astbury took some of the first photos of DNA. Linus Pauling based his entire theory of the structure of DNA on Astbury's photos. Unfortunately, the photos were blurry, and it was still difficult to make out the structure of the molecule.

Levene's nucleotides, Chargaff's rules, and X-ray crystallography would all be critical to discovering the structure of DNA.

The RACE to SOLVE DNA

In 1951 there were three major laboratories working on figuring out the structure of DNA. William Lawrence Bragg headed the Cavendish Laboratory in Cambridge, John Randall headed the King's College lab, and Linus Pauling worked from his lab at the California Institute of Technology. All of them were very experienced and intelligent scientists. Each of them wanted to be the first to crack the code. The race was on.

William Lawrence Bragg and Cavendish

Australian-born British scientist William Lawrence Bragg had been one of the youngest scientists to receive the Nobel Prize at the age of twenty-five with his father, William Henry Bragg. The older Bragg was also a physicist who developed the X-ray spectrometer (a device that make electrons jump around and helps scientists figure out the different elements in each substance). As you might remember, X-ray crystallography is the process of shooting X-rays at crystal structures to make patterns with rays, a technique that was useful for photographing really tiny structures. The first structure to be solved using this method was table salt.

In 1937, Bragg was placed in charge of the laboratory at Cavendish at the University of Cambridge. He divided the laboratory into small groups. Soon after, two scientists by the names of James Watson and Francis Crick would begin to look for the structure of DNA under Bragg's direction. Funnily enough, the chemist Chargaff didn't have much faith in their abilities; he described Crick as having an "incessant falsetto" and Watson as having a "gawky young figure."

John Turton Randall and King's College

Less than a hundred miles away, another team was working on the same problem. The team at King's College in London was headed by John T. Randall, a British physicist who pioneered a way for radios to operate on microwave frequencies during World War II. After the war, Randall became the director of the newly created biophysics research unit at King's College. He recruited Maurice Wilkins, whom he had worked with previously, to study DNA. Rosalind Franklin, an expert on X-ray diffraction, was invited to work with Wilkins about four years later.

Unfortunately, Wilkins and Franklin didn't really work well with each other. This was probably because of a misunderstanding when Franklin was first hired. According to letters, Randall didn't tell Franklin that Wilkins was already working on DNA when he invited her to come to King's College, so she thought that she would be leading the project or doing it on her own. When Franklin and Randall had their first meeting, Wilkins wasn't there (he was on holiday). Wilkins, on the other hand, thought that Rosalind Franklin was supposed to be his assistant.

In response to Wilkins's complaints, Randall told Wilkins that if he had been present at the first meeting, Franklin would have been more accepting of him as a partner in research. He stood by his decision to hire Franklin and wrote, "No one could doubt Franklin's ability." But the division in the King's College lab would continue to plague the duo as they worked to find the structure of DNA.

Linus Pauling

Across the ocean, an American chemist named Linus Pauling was also working on DNA at the California Institute of Technology (Caltech). Pauling had already carved out a reputation for himself for his superb contributions to our knowledge of proteins. Pauling wasn't interested in DNA at first. He did research on inorganic chemistry—the study of the bonds between inorganic (not living) molecules, such as the bonds between salts or metals. However, Warren Weaver, the head of the Rockefeller Foundation's natural science division, believed that biology would soon be the next trendy field of study in science because of new technologies such as X-rays and chromatography (a technique for separating mixtures into separate components). He offered Pauling funding to conduct research on organic chemistry, the chemistry of living things. Pauling shifted from studying inorganic chemistry to the study of proteins, which are one of the most crucial building blocks for our cells (an average human liver cell has approximately eight billion proteins!).

Pauling used his experience with inorganic chemistry to help him. In inorganic chemistry, all molecules, no matter how complex, are built up of simpler, smaller molecules. Pauling applied this rule to organic chemistry, and more specifically, his study of proteins. Pauling first figured out how amino acids (the building blocks of proteins) are structured then figured out that those amino acids often form a spiral shape, called an alpha helix. By figuring out the structure of proteins, scientists could figure out how they worked and what could cause them to malfunction. Pauling's work earned him a lot of admiration from other scientists and some called him "the Wizard of Caltech."

Pauling's signature research method was to use all the information he could find (the amounts of different substances, the angles of the different bonds, etc.) to build detailed scaled models. Watson and Crick would later use Pauling's method to make their discovery.

In 1951, Pauling turned his attention to the question of genes. Because he had studied them for so long and become so intimately acquainted with them, Pauling was a huge fan of proteins and believed, like many others, that genes were

Linus Pauling, known as "the Wizard of Caltech" was among the scientists who raced to figure out the structure of DNA.

stored in proteins, even after the Avery-MacLeod-McCarty experiments proved that genes are actually stored in DNA.

Pauling and Bragg in Cavendish had a friendly rivalry because they often worked on the same types of research. Each wanted to publish results before the other. In the 1920s both Bragg and Pauling were interested in studying large inorganic molecules and Pauling was first to solve the structure. In 1951, Pauling beat Bragg's team again when he published his papers on the structure of proteins. Pauling would publish a paper on DNA before Watson and Crick (who worked in Bragg's lab).

LINUS PAULING and the TRIPLE HELIX

In 1951, when Pauling started becoming interested in DNA, he sent a letter to the King's College team for their X-rays on DNA but was rebuffed. Wilkins wrote a letter back saying that they needed more time. Pauling gave up on DNA for a while, turning his attention back to proteins. During this time, as one of the most respected scientists of the time, Pauling received an invitation to speak in England in 1952. While in England, he had a chance to make another trip to the King's College and ask for the X-ray photographs, but he didn't think that Wilkins would have been willing to share. Little did he know, Franklin had just produced her most clear photographs of DNA, which might have helped him solve the structure very quickly. Later, Wilkins confessed that if Pauling had visited their lab in person, they wouldn't have been able to resist showing him the photographs. "If he had come over in 1952 and visited our lab to look at those X-ray diffraction [photos], I'm sure I couldn't have resisted showing him [everything] we had. Because it was this sort of god-like presence. It would be such an honor to show these things," Wilkins wrote.

In the summer of 1952, Alfred Hershey and Martha Chase found a way to tag DNA in proteins with separate radioactive labels and confirmed Avery's conclusion that DNA contains genetic material. When Pauling heard about their experiments at the International Phage Conference, he realized he had been wrong the whole time: genes weren't located in proteins after all. With that realization, he renewed his efforts in discovering the structure of DNA.

Pauling's Hypothesis

Because Pauling didn't have Franklin's photographs, he had to rely on Astbury's X-rays, which were incredibly blurry. Pauling knew from Astbury's photos and other evidence that the structure had to be something that looked like a helix—a three-dimensional coil that looks like a staircase. He knew that there had to be a phosphate sugar backbone and bases (adenine, thymine, guanine, and cytosine). He decided to put the phosphate backbones into the center of a helix. He thought it would be most likely for the bases to face outward from the backbone of the helix since that would make everything fit together better. It would be much more difficult to cram all the nucleobases into the inside of the helix.

Pauling worked with his assistant Robert Corey on the problem for months and kept running into problems. With his tightly packed structure, he couldn't find a way for the phosphates to stay in the center. Even so, he invited a group of friends to look at his findings and decided that he would go ahead and publish a paper on them.

In December 1952, Pauling published a paper called "Proposed Structure for the Nucleic Acids" in the Proceedings of the National Academy of Sciences. He proposed a triple helix structure with the phosphate and sugar backbone in the center and the nucleobases sticking outward attached to the phosphates. This would be like three spirals intertwined with each other. "The structure really is a beautiful one," he wrote in one letter to a colleague.

However, Pauling still had doubts about his structure. He wrote in his paper that the structure was "extraordinarily tight" and was "probably capable of further refinement."

Pauling's Mistakes

Watson and Crick would later prove that Pauling's structure was wrong. Since phosphate molecules are naturally negatively charged, they would repel each other, and because Pauling put all of his sugar-phosphate backbones in the center of the helix, there would be nothing to hold the structure together. Since the negative phosphates would push each other away, the only way to hold the structure together would be hydrogen bonds (since hydrogen ions have a positive charge). However, it is impossible for phosphates to bond together with hydrogen since they lose their hydrogen in normal pH. Crick wrote to Pauling, "We were very struck by the ingenuity of the structure ... The only doubt I have is that I do not see what holds it together."

Why Pauling Failed

We can only speculate on why Pauling was not the one to discover the structure of DNA. His work "The Nature of the Chemical Bond" was considered one of the most essential texts of the time, and his valance electron theory is still being used today.

Perhaps one of the biggest reasons why Pauling failed was because he simply didn't know enough. While Pauling used many very good X-ray photographs and meticulously accumulated data for his proteins, he didn't have the same knowledge base for DNA. He didn't have access to clear photographs, and he may have been overconfident since he knew that DNA was a simpler molecule than proteins. He also didn't take DNA very seriously until very late in game.

WATSON and CRICK'S ATTEMPT

In early February 1953, Pauling's son, Peter Pauling, who worked in Watson and Crick's lab, showed them a copy of his father's paper. Initially, they were worried that the race to find the structure of DNA was over and Pauling had won. "Well I was scared," Watson said later.

Watson and Crick had begun research on DNA in 1951, about the same time that Pauling had begun dabbling in DNA as well. Watson was a quiet young scientist who had wanted to work at King's College with Wilkins or at Caltech with Pauling but had been turned down at both places. As a plan B, he decided to go to Cambridge so that he could perfect his skills in X-ray crystallography. Crick was a thirty-five-year-old scientist who was known for being talkative. Together, they were less experienced and less decorated than Pauling, but they were

James Watson and Francis Crick accept their Nobel Prizes in 1962.

determined to faithfully copy Pauling's methodology and "beat him at his own game."

Earliest Guesses and Crick's Math Model

Watson and Crick had read a lot of Pauling's work and knew how he thought. They imitated that thought process by trying to figure out the overall structure of DNA first before working out the details. They first guessed that DNA was shaped in a triple helix just like Pauling did. In their initial model, they put the phosphates on the inside and the bases on the outside, but they couldn't get the chemistry to work, for the same reason that it wouldn't work for Pauling (there was too much negative charge in the center).

Proud of their work, Watson and Crick invited Maurice Wilkins and Rosalind Franklin to see their results. But Franklin wasn't very impressed. She pointed out the major flaws with their design: not only was it impossible to have all the phosphates in the center of the helix, their model wouldn't allow DNA to soak up much water. Through her own experiments, Franklin had discovered that DNA soaks up a lot of water, about ten times the amount that their model would allow.

Watson and Crick's boss, Bragg, was so embarrassed by their blunder that he immediately ordered them to stop building DNA models. Crick was to go back to researching proteins, and Watson was forced to study the tobacco mosaic virus. But neither was fazed, and they continued racking their brains for a solution to the DNA problem. Watson feverishly studied up on chemistry. Crick gave Watson a copy of Pauling's famous work, "The Nature of the Chemical Bond," which Watson studied meticulously. "Somewhere in Pauling's masterpiece," Watson said later, "I hoped the real secret would lie."

Confirmation of the Double Helix

Because of their own mistakes, Watson and Crick knew that Pauling's paper on the triple helix was incorrect since it proposed a similar model to the one they had proposed. "If a student had made a similar mistake, he would be thought unfit to benefit from Caltech's chemistry faculty," Watson later wrote arrogantly.

After Watson and Crick had seen Pauling's faulty model, they rushed to tell their supervisor the great news—the race to find the structure of DNA was still on. Bragg gave them permission to start working on DNA again. Crick went to King's College to tell Wilkins and Franklin the good news as well. This visit would turn out to be fateful in their research.

According to Wilkins, Crick first visited Franklin in her lab, but they got into an argument when Crick told Franklin she didn't understand her own data. As Crick left in a hurry to escape the furious Franklin, he bumped into Wilkins in the hallway. Wilkins showed Crick one of Franklin's best X-ray photographs. It would become famously known as photo 51.

Crick had already guessed that DNA was shaped in a double helix. He had done mathematical transformations of helix shapes before and had an idea of what the X-ray diffraction photographs would look like if the structure was a double helix. Franklin's photo, which was the clearest photo of DNA taken at the time, was exactly what Crick was looking for. The first photo that Crick had obtained from Wilkins and Franklin's lab had showed the outside, crystal structure of DNA. This new photo more clearly showed the inside structure. Crick was now sure that DNA was in the shape of a double helix.

Problems with Chemistry

Watson originally hypothesized that the chemical bases could bond with themselves. Adenine could bind with adenine, and thymine could bind with thymine. However, Chargaff's rules stated that the amount of adenine had to equal the amount of thymine and the amount of guanine must equal the amount of cytosine, and Watson's hypothesis didn't explain anything about this.

In addition, with the original hypothesis, not all the molecules could fit in the structure. The base pairs would be unequal sizes, and the backbone would have to shift in and out in a jerky way. Even so, Watson was confident that he had found the answer.

He wrote a letter to his colleague, tearing apart Pauling's paper and stating that he had "a very pretty model, which is so pretty that I am surprised that no-one ever thought of it before." However, when Watson and Crick showed their model to Jerry Donohue, an experienced chemist who worked at Caltech with Pauling, they were told that there were flaws with their model. He told Watson and Crick that the way they had used to fit the bases into the structure were based on textbooks with outdated information. Donahue suggested a new shape for the bases, one where they would link together through hydrogen bonds. With this new information, Watson and Crick returned to the drawing board.

The SOLUTION

Watson worked out the final structure using cardboard cutouts. He discovered that when adenine was paired with thymine and guanine was paired with cytosine, the structure fit perfectly. It also explained Chargaff's rules; since the adenine bonded with thymine and guanine with cytosine, it would only be natural that their quantities would be equal. The final structure was held together by the strength of hundreds of hydrogen bonds. Each nucleobase was attached to another nucleobase with hydrogen atoms (adenine and thymine share two hydrogen atoms and cytosine and guanine share three). Hydrogen bonds are generally a weak type of chemical bond, but when there are hundreds of hydrogen bonds in a structure, the structure will be very stable. "It all happened in 2 hours," Watson later recalled.

How DNA Works and Its Implications

Crick later claimed that they had discovered the "secret of life." Their discovery showed scientists how DNA could be replicated. Since every base has a pair, A with T and G with C, only one side of DNA is needed to make the other half. DNA can split into two pieces and form the missing side up like a puzzle. Enzymes would be able to bond to the bases and help split them up as well as help copy the DNA.

Linus Pauling, when he saw the paper, said that it was "essentially correct." Even Rosalind Franklin agreed that they had come up with the solution. Through her extensive research, she had begun to realize that the structure of DNA was a double helix as well.

Why Watson and Crick Succeeded and Others Didn't

So why did Watson and Crick beat all the other teams to the punch? In King's College, Wilkins and Franklin were the best X-ray crystallographers at the time. Erwin Chargaff and Linus Pauling were excellent chemists.

Maybe it was because they worked really well with each other. Wilkins didn't respect Franklin as a woman. Franklin thought Wilkins was getting in the way of her project. According to Watson, Franklin didn't want to believe DNA was a double helix until very late because Wilkins believed DNA was a double helix. In addition, Franklin refused to build models until she had gathered all the information she could. When Watson and Crick urged Wilkins to build models, he told them that Franklin would leave in two months and he would build models after she left.

Maybe it was because Watson and Crick were willing to take every bit of information they could get. Pauling ignored Chargaff's rules because he thought Chargaff was annoying. Chargaff ignored X-ray crystallography because he didn't understand it and didn't believe that it was very important.

Watson and Crick were able to piece together all the existing information on DNA in a way that no one had been able to do before. They were able to get help from various sources and piece together all the information—Crick's knowledge of X-ray diffraction, Franklin's photograph, Pauling's modeling techniques, Chargaff's chemistry, and Jerry Donohue's advice to finally solve the mystery of DNA.

The CONTROVERSIES

There are several controversies that people talk about when referring to Watson and Crick's paper. Many people believe now that Rosalind Franklin wasn't given enough credit by Watson and Crick. Her photograph was essential to the discovery of the double helix structure, and they might never have been able to publish without her photo. What's more, Wilkins showed Crick Franklin's photograph without her knowledge or approval. Watson's book *The Double Helix* has also received criticism for its portrayal of Franklin. It referred to her disrespectfully as "Rosy" and described her as poorly dressed and disagreeable. Rosalind Franklin never received a Nobel Prize for her contribution because she died of ovarian cancer and was only mentioned in a footnote in Watson and Crick's paper.

Linus Pauling was also unhappy about the amount of credit that Watson and Crick gave to Jerry Donohue, the scientist who had corrected Watson on the way he paired the bases together. "He's the only person whom they thank in their paper on the double-helix," Pauling said. "They could have thanked him even much more strongly, I think, than they did."

Additionally, neither James Watson nor Francis Crick were even initially supposed to be working on the structure of DNA. According to newly discovered letters published in the journal *Nature*, Randall and Bragg had made a gentleman's agreement at the very beginning of the race: Bragg's lab (with Watson and Crick) would work on proteins, and Randall's lab (with Wilkins and Franklin) would work on DNA. When Wilkins learned that Watson and Crick had been secretly working on DNA anyway, he sent a letter to them, asking them to stop. He added a note, trying to smooth over his relationship with Crick, saying, "This is just to say…how rotten I feel about it all and how entirely friendly I am (though it may possibly appear differently). We are really between forces which may grind all of us into little pieces. I had to restrain Randall from writing to Bragg complaining about your behaviour." In response, Crick wrote back, "cheer up and take it from us that even if we kicked you in the pants it was between friends. We hope our burglary will at least produce a united front in your group."

The IMPACTS

Although the discovery didn't have any immediate impact, when the double helix was accepted into the scientific community, its impact was significant. Scientists now agree that the double helix was the single most important discovery in the twentieth century. Watson and Crick's paper created the field of microbiology. Wilkins later took hundreds of X-ray photos of DNA in various different species including squid, chicken, mice, and salmon with increasingly better X-ray technology. After almost ten years, Wilkins was able to fully verify Watson and Crick's paper.

The mystery of genetics was finally solved. After hundreds of years, scientists now knew that genetic information is passed on to offspring from both the mother and the father through a fairly simple molecule. Scientists now knew where genetic information in cells could be found and how cells replicate. They also knew that every gene—from hair color to height—is coded with a four-letter code, made up by the four bases adenine, thymine, guanine, and cytosine. They now knew how DNA could replicate itself over and over again without changing. James Watson and Francis Crick had finally drilled down to the secret of life.

By 1973, researchers at Stanford patented a way to create recombinant DNA (essentially what is known today as genetic modification). This paved the way for the biotech industry, genetic modification, new drug development, vaccine development, cloning, and much, much more.

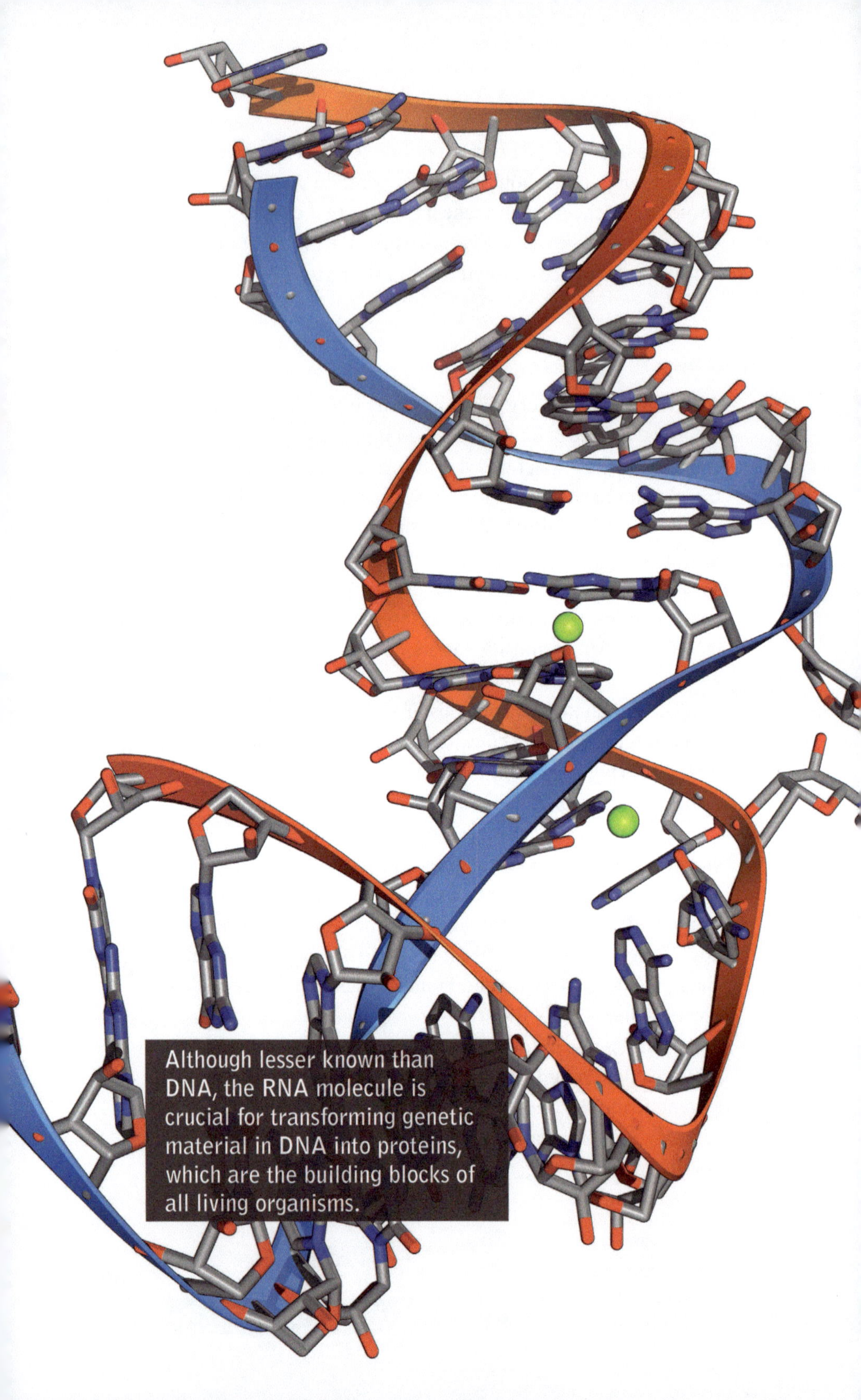

Although lesser known than DNA, the RNA molecule is crucial for transforming genetic material in DNA into proteins, which are the building blocks of all living organisms.

The DISCOVERY of RNA

Up until this point, we haven't talked much about another molecule very similar to DNA, called RNA. RNA is a single-stranded molecule. Like DNA, it also has four nucleobases — adenine, uracil, cytosine, and guanine. Unlike DNA however, RNA doesn't form the double helix structure like DNA does. Although there isn't much documentation about when scientists knew about the existence of RNA, most people date the discovery of RNA back to Friedrich Miescher (whom we first discussed in chapter 1). Various scientists then figured out that RNA was very similar to DNA and figured out the nucleobases in RNA. But it wasn't until after the discovery of the DNA double helix that scientists really turned their attention to RNA.

After identifying the structure of DNA, Watson and Crick realized that they needed to figure out what DNA actually did. They knew that DNA had something to do with RNA and RNA had something to do with proteins, but they didn't know how exactly they related. Watson reached out to Russian physicist George Gamow and together they formed a gentleman's club called the RNA Tie Club. The club consisted of twenty world-class scientists who all wore woolen ties with an embroidered helix on them. Gamow used math to figure out that only three nucleobases were needed to code all twenty amino acids. Later, biologist Sydney Brenner, who was also in the club, came up with the idea of a codon. A codon is a group of three nucleotides that would code for a specific amino acid, in keeping with Gamow's theories.

Next, Crick proposed something called the adaptor hypothesis, which proposed that different messenger RNA molecules move the twenty amino acids into order to form proteins. This hypothesis was later confirmed by Mahlon Hoagland and Paul Zamecnik in 1956, when they discovered the adaptor molecule. It was called transfer RNA, or tRNA.

In 1958, Crick published a paper called "On Protein Synthesis." This paper laid out how the sequence of bases determined the amino acid sequence that determined the 3D structure of proteins.

By around 1960 they found out that there are three types of RNA. Messenger RNA are large groups of RNA molecules that transcribe the information provided in DNA. They do this through a process called transcription. An enzyme called RNA polymerase comes in and unzips the DNA double helix. Then, it matches RNA bases to the DNA bases (guanine matches with cytosine, adenine matches with uracil). Once messenger RNA has gotten the genetic information, it moves out of the nucleus of the cell and into the cytoplasm, where a process called translation takes place. Molecules called transfer RNA each have three nucleobases on one end and an amino acid on the other end. The three nucleobases match up with the nucleobases of the messenger RNA. As transfer RNAs continue attaching to the messenger RNA, the amino acids on the other end link up. Yet another type of RNA—ribosomal RNA—helps this process along. Once the whole process is complete, the amino acids detach and fold in on themselves to create a protein.

In 1961, Marshall Nirenberg and Johann H. Matthaei, American scientists not in the RNA Tie Club, discovered the first amino acid, called amino acid F, that was coded by DNA. They were able to isolate a few nucleotides at a time and see what amino acids they created. Amino acid F is also known as phenylalanine, which is created from one long chain of uracil. Synthetic phenylalanine is now commonly found in chewing gum.

By 1966, "It was all over," according to Watson. In 1970, Watson published a paper called "Central Dogma of Molecular Biology," which illustrated the progression of genetic information from DNA to RNA to amino acids.

James Watson was twenty-five years old when he helped discover the structure of DNA.

DISCOVERING the SECRET to LIFE

Watson, Crick, and Wilkins would win a Nobel Prize in 1962 for their contributions to medicine. Both Watson and Crick would go on to be remembered in textbooks and popular media as the scientists responsible for one of the most important scientific discoveries of the twentieth century. But as you know by now, Watson and Crick didn't work alone. They had a lot of help from scientists like Franklin, Pauling, and Donahue, and from the work of scientists many years before.

Over the course of the last few chapters, you saw scientists begin with a general idea of how traits were passed down from parent to child (in Mendelian genetics) to a more detailed understanding of how cells divide and replicate genetic information. With the discovery of the structure of DNA, scientists were finally able to break down the process of inheritance to the smallest level. Knowing the structure of the DNA molecule allowed scientists to understand how DNA functions in a way they had never been able to understand before. This discovery would also pave the way for more research into how RNA works and for important technologies, which we will discuss in the next chapter.

Model Building

Model building is the technique that Watson and Crick used to ultimately figure out the structure of DNA. Model building started out sometime around 1860 with the German chemist August Wilhelm von Hofmann, who built a model of methane. At the time, no scientists built models because most of them didn't believe that atoms existed. They didn't believe that atoms could be three-dimensional particles arranged in complex structures.

In 1873, Jacobus van 't Hoff and Achilles Le Bel made an important discovery that showed scientists that molecules are created from three-dimensional atoms in specific structures. At this point in time scientists were puzzling over the phenomenon of isomers, substances that had the same chemical composition but different properties. For example, a propanol molecule is made of three carbon atoms, eight hydrogen atoms, and one oxygen atom. A methoxyethane molecule has the same exact chemical composition. Van 't Hoff and Le Bel discovered the structure of carbon bonds and showed that they were important in determining the properties of a substance. Propanol and methoxyethane both have the same atoms but the hydrogen and oxygen are bonded to the carbon atoms in different ways. That's why they behave differently. The two molecules are considered isomers.

Together, these scientists established a field called structural stereochemistry, which is the study of the structure of molecule using known distances between atoms. Aside from Watson and Crick, Linus Pauling was another scientist famous for his use of stereochemistry.

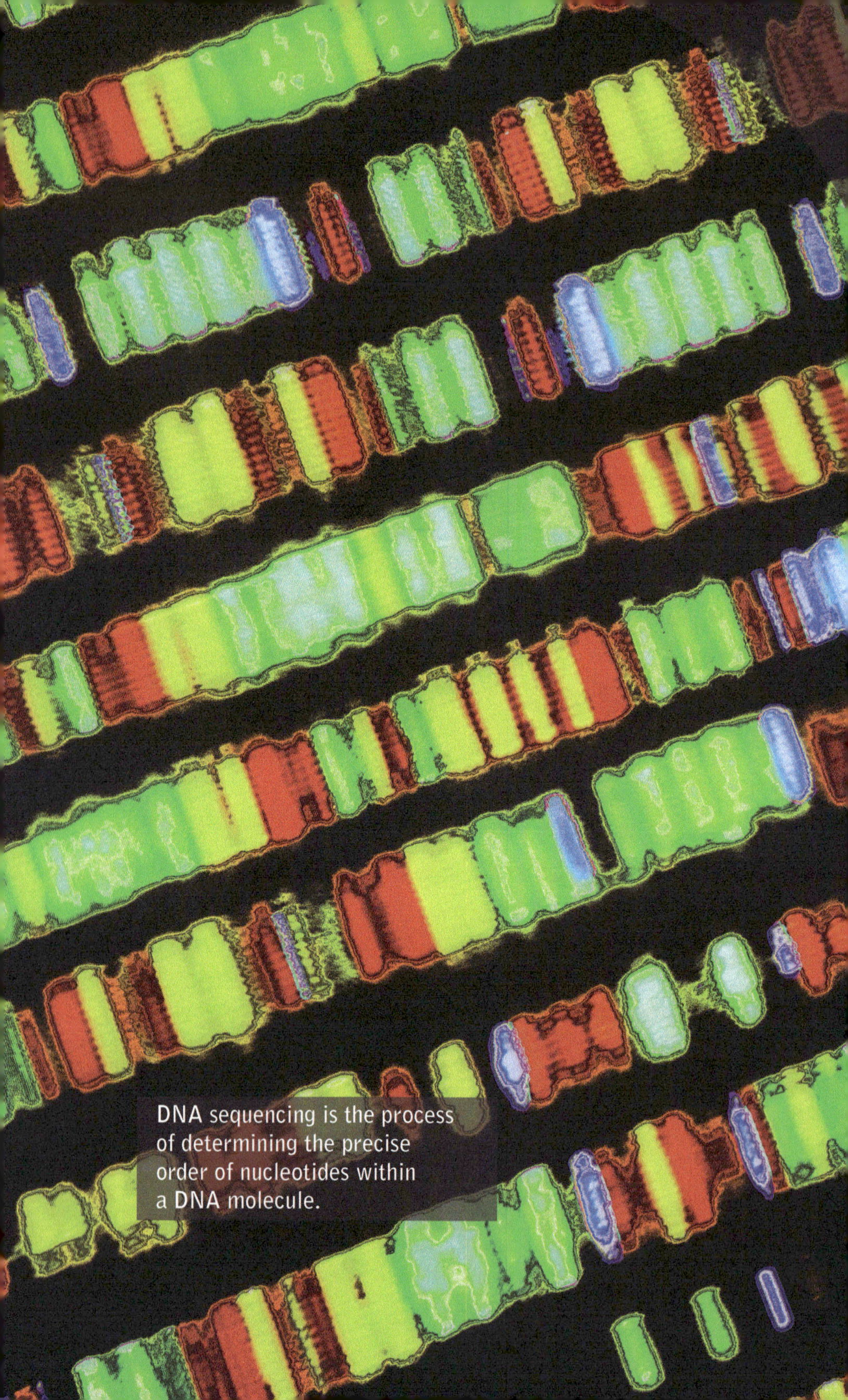

DNA sequencing is the process of determining the precise order of nucleotides within a DNA molecule.

CHAPTER 5

The Influence of DNA and RNA Today

The discovery of DNA and RNA changed medicine forever. More than a hundred years after Miescher first isolated nuclein in the lab, scientists have mapped out all thirty thousand genes in the human body, used genetic testing to better understand how genes play a role in diseases, and even modified genes in bacteria, animals, and plants. Nonc of this would have been possible without Miescher's discovery. Here is a look at some of the applications that DNA has today.

"The MOST WONDROUS MAP PRODUCED by HUMANKIND"

Imagine if you never saw a map of a new neighborhood. You would know there were streets out there, but you wouldn't know exactly how many or how they intersect to get to the main road. You wouldn't know which streets were most important to the neighborhood, and which ones led nowhere. Without that bird's-eye view, you would get lost pretty quickly.

That's how scientists felt about DNA. By the 1990s, researchers had a pretty good idea of what DNA was and how powerful and important it was in shaping who we are, but they didn't know exactly what those genetic pairs were or where they stood in relationship to each other.

That's why, in 1990, the United States government decided to fund something called the Human Genome Project—a project that would map out all thirty thousand genes in the human **genome** (a genome is the entire set of DNA in an organism's body). They believed that by mapping out all the individual nucleotides in the human genome, they would be better able to figure out when, where, and why genetic mutations occur and how they impact human health.

To create the Human Genome Project, scientists from institutions all over the world, including Germany, China, France, Japan, and the UK, took blood samples from anonymous volunteers. They then read the DNA in those blood samples using a lengthy process called BAC-based sequencing.

What Is BAC-Based Sequencing?

BAC-based sequencing was first developed in the 1980s. Because DNA is so small (a strand of DNA containing 2.5 million base pairs is only about 5 centimeters [about 2 inches] long...smaller than your pinky!), it's impossible to "read" the base pairs of DNA with the naked eye. The BAC-based sequencing process allows scientists to read all the base pairs in a strand of DNA with the help of bacteria.

To start the process, scientists take several copies of the human genome from various anonymous volunteers. They separate the strands of DNA by putting it in a high temperature environment (the process is called denaturing). Those strands of DNA, which contain over 2.5 million base pairs, are then cut into pieces about 150,000 base pairs long (much more manageable!). Those pieces are then inserted into a piece of bacteria, and as that bacteria replicates, it also replicates pieces of that DNA.

By this point, scientists have several copies of pieces of DNA, but they still have no idea what base pairs are in each

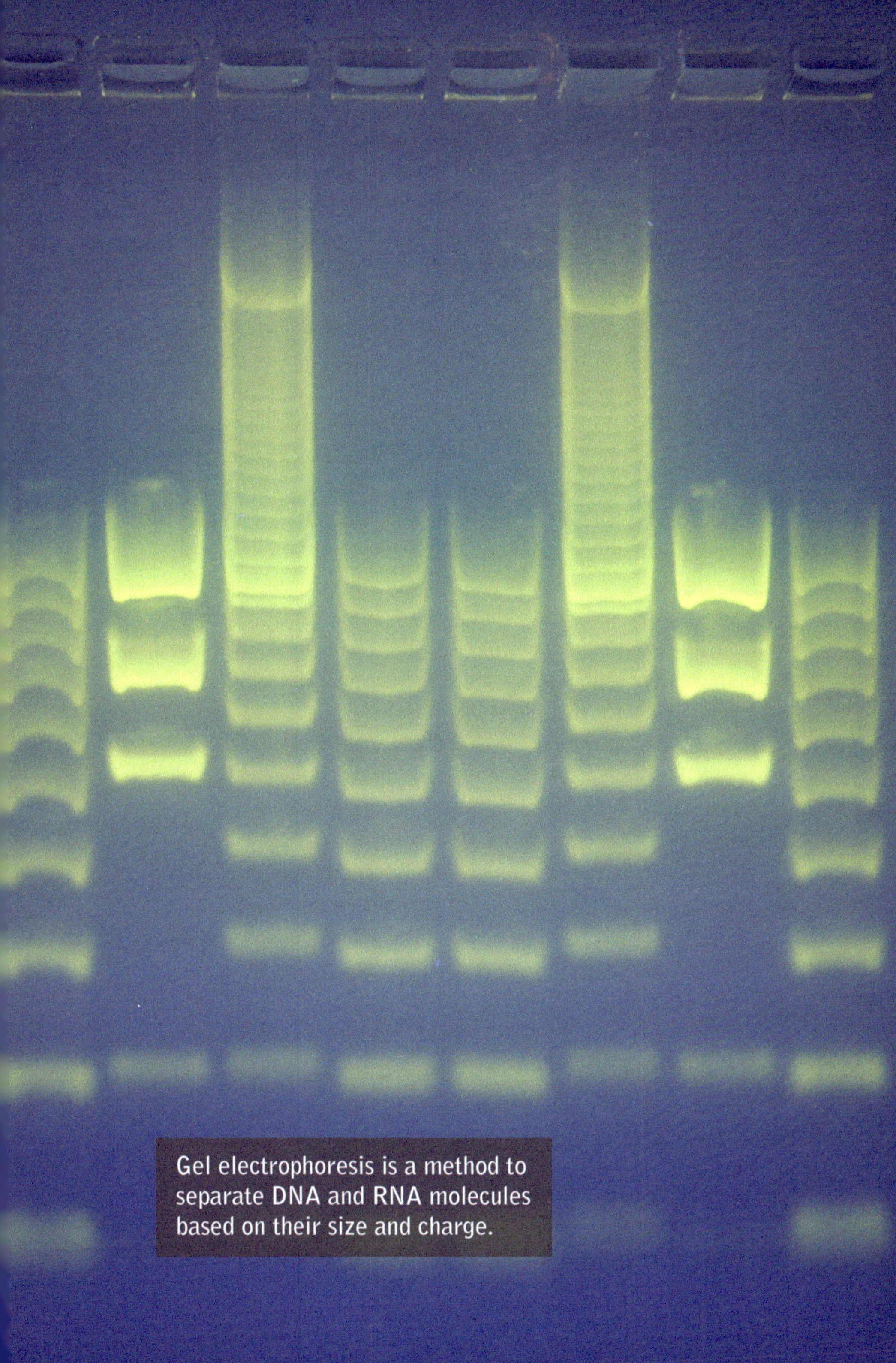

Gel electrophoresis is a method to separate DNA and RNA molecules based on their size and charge.

of these fragments. To figure that out, scientists have to do a little detective work. Remember that each DNA base pair—adenine, cytosine, guanine, and thymine—pairs only with its opposite (adenine with thymine, and cytosine with guanine). Scientists take a bunch of free-floating base pairs and pick one type (let's say adenine in this example) and mark it with a color. They then let those free-floating base pairs loose to pair up with the DNA base pairs in the fragment. Because scientists have marked one of the base pairs, in this case, adenine, the matching-up process will stop at adenine. And because they know that adenine only pairs with thymine, they know that the last DNA base pair of the original fragment is thymine!

Researchers repeat this process hundreds and thousands of times with different fragments and different base pairs until they figure out the name of each base pair in our genome. It's a time-consuming process, but luckily, different teams of scientists from around the country all took responsibility for a small part of the human genome during the international Human Genome Project. Together, those scientists were able to piece together all the information and get a map of the human genome.

When the Human Genome Project was completed in 2003, scientists finally had a complete copy of the genetic instructions that come preprogrammed in each human being's body. President Bill Clinton called it, "without a doubt… the most important, most wondrous map ever produced by humankind," during his remarks at the White House when he unveiled the first draft of the human genome in 2000.

GENES and MEDICINE

With a map of the entire human genome in hand, scientists set about figuring out how changes in that genome might impact human health. Although all humans share 99.9 percent of the same DNA, the remaining .1 percent affects how we metabolize drugs and pass diseases on to our offspring. In

2002, for example, researchers found that certain patients who had a gene called HLA-B*57:01 had a 60 percent chance of developing a hypersensitivity reaction when they took the HIV drug Abacavir. Before scientists knew this, doctors had to use trial and error to figure out which patients would react badly to Abacavir. Now, doctors just have to order a blood test before prescribing the drug.

Precision Medicine

The case of Abacavir was the just beginning of a burgeoning field of research called **precision medicine**. Precision medicine, also known as personalized medicine, allows scientists to look at how a specific individual might react to a certain drug or the chances of developing a disease based on his or her specific genetic makeup. In 2013, President Barack Obama announced the Precision Medicine Initiative, which called for $215 million to build a national, large-scale research participant group in collaboration with the National Institutes of Health (NIH). The initiative also helps the National Cancer Institute research genetic causes of cancer. The NIH research group will include about one million Americans who will volunteer information about their DNA, behavioral data, health records, and other medical information, which will then be studied by researchers. Scientists hope that access to this data will help them better observe the effects of drugs and medical devices.

A White House statement introducing the Precision Medicine Initiative said:

> ... [I]nformation from the cohort will be a broad, powerful resource for researchers working on a variety of important health questions. The program will seek to extend precision medicine's success to many diseases, including common diseases such as diabetes, heart disease, Alzheimer's, obesity, and mental illnesses like

depression, bipolar disorder, and schizophrenia, as well as rare diseases. Importantly, the cohort will focus not just on disease, but also on ways to increase an individual's chances of remaining healthy throughout life.

To aid in the development of precision medicine, scientists have built upon the Human Genome Project and have created several additional gene libraries. In 2005, scientists published the International HapMap, a catalog that records common genetic variations in human beings and helps researchers link those variations to diseases and treatment responses. The name "HapMap" is short for haplotype map. Haplotypes are specific groups of genes that are inherited together from the parents. The HapMap identifies those groups of genes and what they code for. The same year that researchers published the HapMap, other researchers from institutions worldwide created the Cancer Genome Atlas, a project aimed at characterizing all genomic changes involved in human cancers. The project ultimately mapped out more than twenty different cancers, including cervical cancer, esophageal cancer, breast cancer, and liver cancer.

Genomics has already helped scientists develop some of the newest cancer treatments. For example, the leukemia drug Gleevec was created specifically to block an altered enzyme produced by a fused version of two genes found in chronic myelogenous leukemia, and the breast cancer drug Herceptin was developed specifically to help women whose tumors have a particular genetic profile called HER-2 positive. Scientists have also used their knowledge of genetics to better understand which drugs to prescribe certain patients, even if they're exhibiting the same disease. For example, researchers discovered that lung cancer patients that have certain mutations respond well to the drugs Iressa and Tarceva, while certain other drugs are not helpful for colon cancer patients that have a mutation in a gene called KRAS.

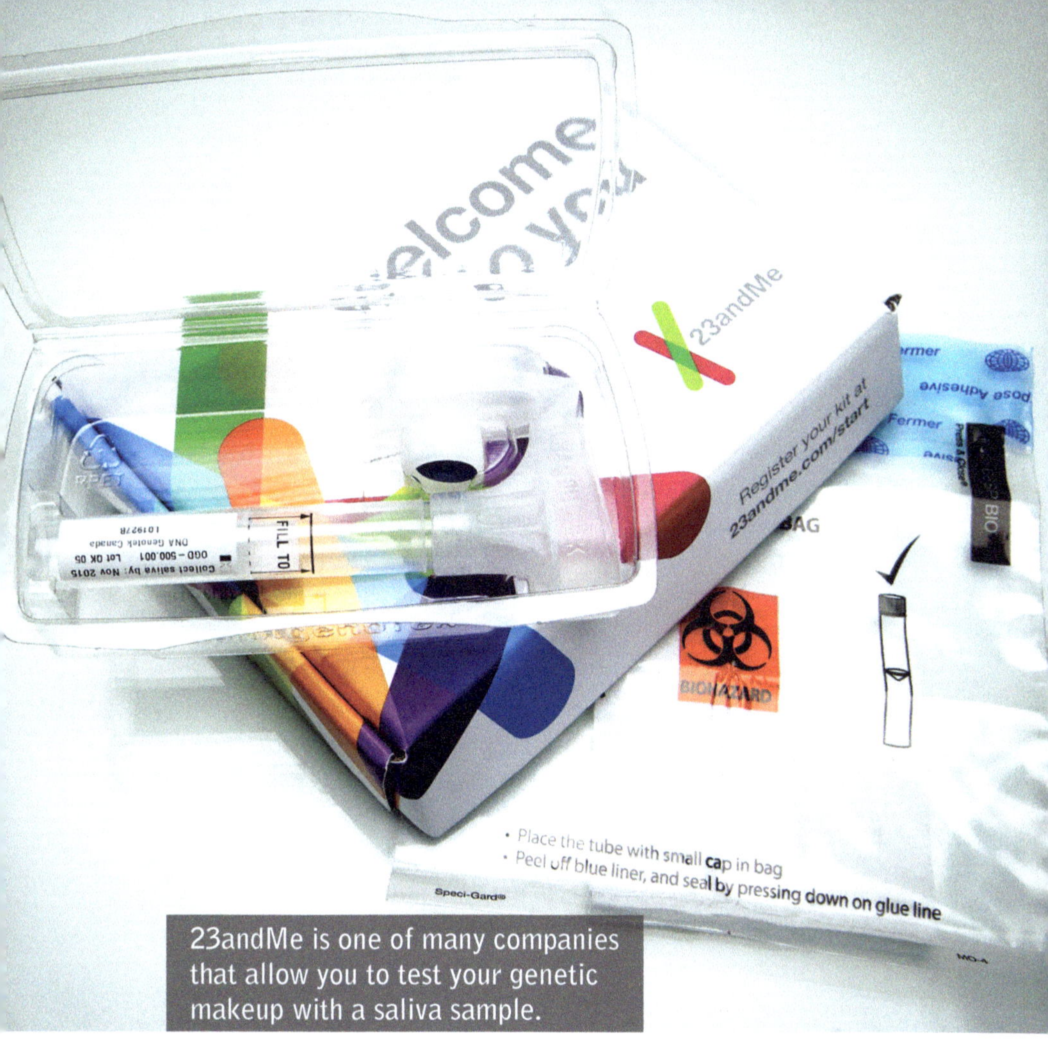

23andMe is one of many companies that allow you to test your genetic makeup with a saliva sample.

DIY Genetic Testing

Companies like 23andMe and GeneDx now offer do-it-yourself gene-testing kits that allow everyday people to test their DNA. Customers mail a cheek swab or saliva sample to a company, and their saliva is analyzed for some 650,000 different genetic variations that may be linked to certain diseases such as celiac disease, breast cancer, and multiple sclerosis. Customers then receive a report with results that discuss their percentage of risk for these diseases according to their genetic profile.

In 2015, the Food and Drug Administration (FDA) approved the first direct-to-consumer genetic testing kit on the market, after withholding approval for many years (the kit is manufactured by 23andMe). This industry has exploded in popularity in the last couple of years: according to some estimates, the DIY genetic-testing market will reach more than $230 million by 2018.

However, many of these companies have received criticism from the public. While proponents of do-it-yourself DNA testing argue that these kits help patients take charge of their own health, critics worry that these kits can cause unnecessary anxiety and alarm. After all, diseases can be caused by many different factors, including lifestyle, and just because you have a genetic predisposition to a disease, doesn't necessarily mean that you're going to develop it. In addition, many of these tests haven't actually been proven to accurately predict disease.

A March 2008 article in *Science* magazine analyzed a do-it-yourself test kit that was supposed to test how effective a certain class of antidepressants would be for patients. The companies that made these tests argued that the tests would help doctors make better decisions about which antidepressants to prescribe to patients and at what doses. But clinical studies showed that having access to the genetic information didn't make any difference on whether patients' conditions improved or not. The study authors said that the results were troubling and might cause patients without proper medical training to misinterpret the results. "For example, a patient informed of his or her cytochrome P450 (*CYP450*) profile might independently change the dose of antidepressant medication with adverse health outcomes," the authors concluded in their study.

These concerns may be important to keep in mind since the results of DIY tests can potentially lead to important health decisions. One example of this is the actress Angelina Jolie, who decided to have double mastectomy after using a test that indicated she had a high risk of developing breast cancer (87 percent risk of developing breast cancer and 50 percent risk for ovarian cancer).

Keeping Genetic Information Private

Of course, with all the genetic information now floating around in cyberspace as doctors and researchers send databases to each other, there has been concern about genetic information privacy. People are worried that hackers can easily access genetic information, and many research participants may decline to participate after they hear about the process of donating their genetic information.

A study conducted by researchers at the Whitehead Institute for Biomedical Research in Cambridge, Massachusetts, in 2013 suggested that it was possible to figure out the identity of some of the men who had participated in the Human Genome Project. They wrote a computer algorithm that would guess the rest of a person's gene based on the bits they had in the Human Genome Project database, then matched that information up with genealogical databases. Others have pointed out that the men they used had an unusual amount of genetic information sequenced for various different projects and that it wouldn't be possible to figure out an ordinary person's identity by the little information they had provided to the Human Genome Project. Still, the possibility is there, and it increases as more genetic information is made available.

In response to some of these findings, the NIH has moved some of its genetic information behind password-protected firewalls. Still, others argue that this might make it very difficult for scientists from different organizations to access the information efficiently to do their research. A big part of the Precision Medicine Initiative is to create a secured database for genetic information, although a solution still hasn't been figured out.

There have also been a lot of new laws protecting people from genetic discrimination. The Genetic Discrimination Act of 2008 protects Americans from being discriminated against by health insurance companies and by their employers based on genetic information they have provided for research. This means that health insurance companies can't look for genetic diseases in genetic information and charge someone more because of it. It also means that employers can't make hiring decisions based on whether or not someone has a genetic disease. Between 2008 and 2013, more than a thousand charges of genetic discrimination were filed against health insurance companies or employers.

Reverse Transcription

In 1970, scientists found a process that did the exact opposite of what Francis Crick proposed in his paper "Central Dogma of Molecular Biology." As we saw in chapter 4, normally RNA makes a copy of the DNA code to make proteins. In a process called reverse transcription, scientists copy RNA into single-stranded DNA with an enzyme called reverse transcriptase. A lot of retroviruses use this process naturally to insert their viral RNA into the host's DNA. An example of a retrovirus is HIV.

The discovery of reverse transcription was important not only in figuring out retroviral drugs that suppress HIV, but also in producing insulin for patients with diabetes. Scientists would put a piece of RNA that coded for the production of insulin and the reverse transcriptase enzyme into a bacteria. That bacteria would then copy the RNA into its DNA and as a result, produce large amounts of insulin very quickly.

Gene Silencing

In 1998, scientists found something called RNA interference. Two types of small RNAs called siRNA, or small interfering RNA, and miRNA, or microRNA, bind to mRNA molecules and disable certain parts of the genes. Although a lot of research still needs to be done on RNA interference, many scientists are exploring ways to use gene silencing techniques for everything from insecticides to cancer cures.

MODIFYING GENES for FOOD, HEALTH, and PROFIT

In the spring of 1973, two men peered through a microscope on the campus of Stanford University. Herbert Boyer was a round-faced, curly-haired former activist originally from Pennsylvania, and Stanley Cohen, the son of tailors, was a native of Perth Amboy, New Jersey. They had met the year before at a U.S.-Japan joint meeting on plasmids and instantly connected over their interest in what genes were present in bacteria and how they were arranged.

Years earlier, a scientist by the name of Paul Berg had found a way to splice a bit of the DNA of a bacterial virus in the DNA of a virus naturally found in monkeys.

Boyer and Cohen took the process one step further: they were able to insert genes from a South African clawed frog into the bacteria *E. coli*. It was the first time scientists had inserted genes from one species into another species' genome, and it opened the door for the years of genetic modification to come.

Since Boyer and Cohen's experiments, the field of interspecies genetic modification, often called **transgenic** organisms, has been growing rapidly. Scientists have used genetic modification to create insulin-producing bacteria for diabetic patients, glow-in-the-dark fish, and even more muscular pigs. Researchers have also developed animals to help in environmental efforts. For example, in 2012, a lab at University of Exeter in England began developing a genetically modified zebra fish to detect which parts of fish are affected by environmental pollution. By inserting the gene that makes jellyfish glow green (called green fluorescent proteins) to make the affected parts of the zebra fish glow, researchers now know that many environmental pollutants affect more of the animals' body systems than they previously thought.

Genetically Modified Foods

Scientists have also turned to genetic modification in an attempt to affect the food we produce by making the food tastier and more nutritious, or by making it easier to grow. One of the biggest challenges farmers face in food production is food waste due to plant disease and pests. As a result, scientists have started looking for ways to use genetics to make crops resistant to those diseases. In 1998, the first papaya genetically modified to resist the ringspot virus (which was destroying almost all of Hawaii's papaya crop) was approved for sale on the market. In February 2015, the U.S. Department of Agriculture (USDA) approved the first genetically modified apple that would silence the gene that causes the fruit to brown.

The two most common genetically modified foods in the U.S. food system are soy and corn; 95 percent of soy and 84 percent of corn grown in the United States is genetically modified to resist insects and herbicides. The reason why herbicide-resistant crops are in such high demand is because of a weed killer called Roundup. Farmers use Roundup to kill weeds by destroying a group of amino acids found only in plants (phenylalanine, tyrosine, and tryptophan). The problem is that these herbicides can also kill the crop itself. As a solution, scientists introduced a gene from the bacteria *Agrobacterium tumefaciens* to make them resistant to Roundup. Thus, farmers can use the herbicide without worrying about destroying their crops as well.

Scientists have been looking for ways to change genes in the animals we eat as well. For example, in 2006, scientists introduced a roundworm gene into pigs to produce omega-3 fatty acids, a nutrient more commonly found in fish than in other meats. Researchers have also been experimenting with

More than 80 percent of corn grown in the United States today is genetically modified.

ways for cows to produce lactose-free milk and chickens that are resistant to bird flu. In November 2015, the FDA approved the first genetically modified salmon, created by the company AquaBounty. Scientists inserted a growth gene from the ocean pout, an eel-like creature, to make the salmon grow to market weight in eighteen to twenty months, compared with twenty-eight to thirty-six months for conventionally farmed salmon.

While genetically modified foods appear to offer solutions to many problems, critics argue that they could potentially cause catastrophic harm to the environment and human health. Firstly, by creating plants that are resistant to herbicides, it has encouraged the agricultural industry to apply increasing amounts of herbicides on plants. In the United States, the use of herbicides has increased by an alarming amount—from .4 million kilograms in 1974 to 113 million kg in 2014. Around the world, the use of herbicides has increased ten-fold. As a result, new breeds of weeds that are resistant to normal herbicides have emerged in thirty-six states in the United States. Those weeds can only be killed by extremely toxic herbicides, including 2,4-D, (which is a chemical found in the toxic gas Agent Orange, which was used during the Vietnam War). Many of these herbicides have been labeled by the International Agency for Research on Cancer as possible carcinogens.

In addition, many are concerned that genetically modified animals that have been bred to be bigger and stronger could escape into the wild and wreak havoc on the ecosystem. In the case of AquaBounty salmon, critics worry that if the bigger fish were to escape, they would outcompete wild salmon for food or mates.

In addition, some consumer groups worry that genetically modified foods may have long-term impacts on human health. The National Academy of Sciences reviewed the safety of genetically modified, or GM, crops in 2000 and 2004 and noted that based on lab testing, genetically modified crops don't appear to pose any unique hazards to human health. However, those tests were not directly on humans. The authors also

The genetically modified AquAdvantage salmon can grow to market size in sixteen to eighteen months rather than three years.

noted that genetic transformation has the potential to produce unanticipated allergens or toxins but that more information was needed.

HUMAN GENE THERAPY

About four thousand human diseases have been traced to genetic causes. Many of those diseases may sound familiar to you: cancer, AIDS, cystic fibrosis, Lou Gehrig's disease, and Alzheimer's. Some of these diseases, like cystic fibrosis, are caused directly by the mutation of a single gene while other diseases, like arthritis, are cause by a combination of genetic and environmental factors.

Because genes are so important to human disease, scientists have been exploring ways to use genetic modification to cure disease. Gene therapy treats disease by "repairing" dysfunctional genes or by providing copies of missing genes.

In order to repair genes in the human body, scientists isolate normal DNA and put it in a vector, which is a DNA molecule used as a vehicle to artificially carry foreign genetic material into another cell. Scientists then infect a target cell with the vector. The vector unloads its DNA cargo, which then begins producing the proper proteins and restores the cell to normal. However, there is a lot of risk involved. If the DNA is inserted in the wrong place in the genome, it can cause cancer.

There are two types of human gene therapy: **somatic gene therapy** and **germline gene therapy**. In somatic gene therapy, scientists transfer genes into body cells (like the cells in your heart or liver). In this case, the new genetic material only affects the patient and will not be passed on to the patient's children. In 1992, scientists conducted somatic gene therapy on a twenty-nine-year-old woman to treat a disease called familial hypercholesterolemia, which affects the liver's regulation of cholesterol in the blood. Scientists are now looking at ways to use somatic gene therapy to treat cystic fibrosis, a genetic disorder that damages the lungs and digestive systems.

In germline gene therapy, scientists modify the germ cells (which are either egg or sperm cells, capable of creating an embryo). Germline gene therapy changes the genetic makeup of the egg or sperm of the patient and thus, the new genetic material will be passed on to children and future generations. This has the potential to remove a genetic disease from the family line forever. However, if something goes wrong, it could also affect the family gene pool forever. Currently, germline therapy is illegal in most of Europe.

Besides using gene therapy to simply repair or replace genes, researchers are also looking at ways that gene therapy can deliver genes that will destroy cancer cells or signal cancer cells to revert back to normal cells and deliver viral or bacterial genes to vaccinate people against certain diseases.

CRISPR Technology

One of the newest and most exciting technologies involving genetic modification is called CRISPR. CRISPR stands for "clustered regularly-interspaced short palindromic repeats." The technique allows scientists to cut pieces of DNA more precisely

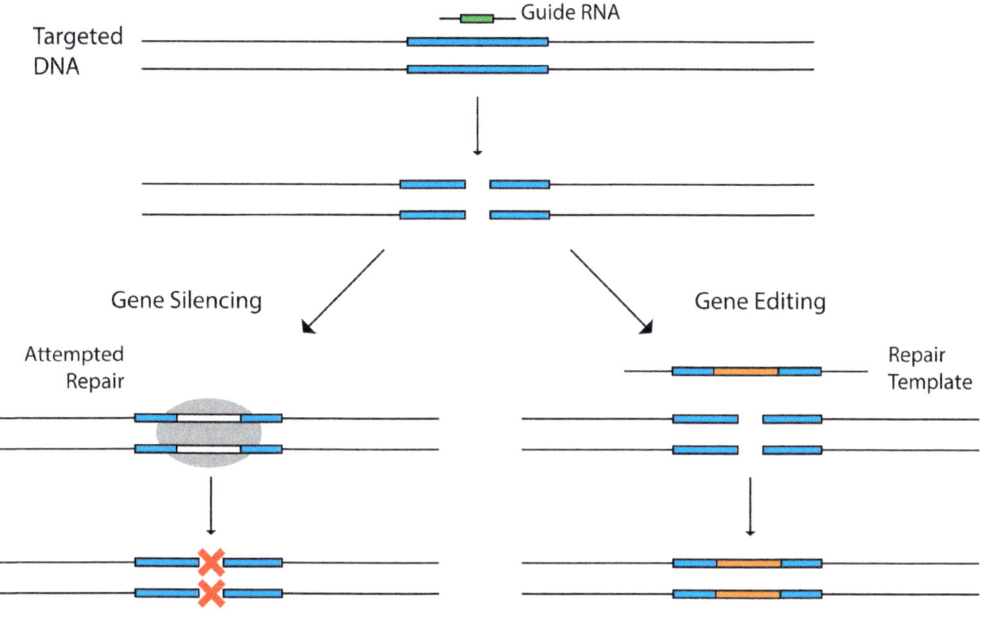

CRISPR technology takes advantage of a protein called CAS9 that cuts DNA.

and quickly than ever before. It is based off of a natural process that bacteria use to protect themselves from invading viruses. When a bacterium detects an invading virus, it will produce two strands of RNA that are identical to a section of one strand of the virus DNA—after the virus injects its DNA into the bacteria. The two strands of RNA form together with a protein called CAS9. CAS9 is a type of enzyme that can cut DNA. When the RNA finds its target and binds with its complementary section on the DNA, the CAS9 cuts the DNA at the beginning and end of that section. Over the past few years, researchers have discovered that they can engineer the process to cut not just viral DNA, but all sorts of DNA across all species. They just have to change the RNA strand to precisely match the section of DNA they want to cut. This process can even be engineered to take place in living cells! In 2015, the journal *Science* named

CRISPR one of the breakthrough discoveries of the year. Of course, the technology has its critics. Many researchers believe that new methods for editing genes aren't necessarily a good thing and they worry more than ever that these methods will be used to alter human genes in a way that is unethical.

The END

Hopefully, after reading the whole story of the discovery of the structure of DNA, you've also gained an understanding that a scientific breakthrough doesn't happen because of one or two people: it's a monumental effort that takes hundreds of years and the combined efforts of many smart people.

The saga of the discovery of DNA finally came to an end in 1953, but that doesn't mean that the quest for more information has ended. There are still more questions: How can we use our knowledge to cure diseases and disable viruses? What are we changing when we change the genetic makeup of plants, animals, and humans? In other words, what do we not know yet? Those mysteries will require a whole new generation of scientists to solve.

The Ethics of Gene Editing on Humans

With our growing knowledge of gene editing, it won't be long before scientists will be able to directly edit the genes of human beings. This worries a lot of scientists, and in December 2015, a group of them convened at the International Summit on Human Gene Editing in Washington, DC, to discuss the ethics of editing human DNA. While gene editing could solve a lot of problems for humans—preventing genetic diseases such as Huntington's disease, for example—many worry that using genetic-engineering techniques to edit genes in human eggs, sperm, or embryos—essentially changing babies before they're even born—could be ethically very problematic. Many people are worried that this could give rise to "designer babies," breeding babies for certain characteristics like intelligence, height, etc. Others worry that a misstep in changing human DNA could have huge ramifications because those changes would get passed down for generations. Scientists like Jennifer Doudna, one of the inventors of the CRISPR technology, warned in her September 2015 TED Talk that scientists need to be careful about the way they use these technologies. "We've called for a global pause for any clinical applications of the CRISPR technology," she said.

Chronology

1820s Gregor Mendel discovers his laws of inheritance

1869 Friedrich Miescher first isolates DNA

1876 Oskar Hertwig first observes meiosis

1882 Walther Flemming publishes book describing mitosis

1909 Phoebus Levene discovers ribose

1929 Levene discovers deoxyribose

1944 Erwin Schrodinger publishes his book *What Is Life?*

1944 Avery-MacLeod-McCarty experiment shows that DNA holds genetic material

1949 Erwin Chargaff announces Chargaff's rules

1950 Rosalind Franklin is invited to work in the King's College Laboratory on DNA

1951 Linus Pauling, Francis Crick, and James Watson begin their search for the structure of DNA

1952 Linus Pauling publishes his proposed structure for a triple helix in the Proceedings of the National Academy of Sciences

1953 Watson and Crick's paper is published in the journal *Nature*

1954 Watson and George Gamow form the RNA Tie Club

1956 The "adapter molecule" is discovered, later known as tRNA.

1958 Crick publishes a paper about the relationship between DNA, RNA, and proteins

1960 The three forms of RNA are discovered: tRNA, mRNA, and rRNA

1961 Marshall Nirenberg and Johann H. Matthaei are first to decode the specific nucleobase order that codes for specific amino acids

1970 Watson publishes a paper, "Central Dogma of Molecular Biology"

2003 The Human Genome Project is completed

2013 President Obama announces the Precision Medicine Initiative

Glossary

allele One specific form of gene; humans usually have two for each trait.

atom The smallest particle of matter that cannot be divided anymore; every atom is made of a single element.

bacteria A type of single-celled organism.

chromatid One copy of a chromosome.

chromatin Unorganized strands of chromosomes before meiosis or mitosis.

chromatography A method for separating a mixture by moving it through a material.

chromosome Threadlike bundle containing a single strand of DNA and protein found in the nucleus of plant and animal cells.

DNA Stands for deoxyribonucleic acid; a molecule in all living cells that carries genes.

enzyme Something that speeds up chemical reactions.

epigenetics Aristotle's theories that organisms are created unformed and then develop into form.

eugenics A belief that the human race can be improved by improving the genetic quality of the human population.

gene Part of DNA that codes for specific traits.

genome An organism's complete set of DNA.

germline gene therapy The practice of modifying genes in germline cells, which are sperm and egg cells.

helix A spiral shape.

hereditary Describes a trait that can be passed on from parent to offspring.

leucocyte A type of blood cell involved in the immune response to foreign substances.

meiosis The process through which sperm and egg cells divide.

mitosis The process through which body cells divide.

molecule A group of atoms bonded together.

nucleic acid The overall name for DNA and RNA.

nucleobase The molecules adenine, thymine, guanine, cytosine and uracil are the nucleobases.

nucleotide A set of three molecules, a phosphate group, a sugar group, and one nucleobase, that make up the building blocks of DNA and RNA.

nucleus An organelle found at the center of the cell, where DNA and some RNA are found.

pangenesis A theory proposed by Charles Darwin that each cell contains seeds that concentrate in the sexual reproductive organs and then get passed on to offspring.

phage A virus that infects bacteria; short for bacteriophage.

precision medicine A method of care that believes in customizing health care to each patient, sometimes through genetics.

protease An enzyme that breaks down proteins.

protein Molecule that performs a variety of functions, including jump-starting reactions, building muscle, and more; proteins are made up of chains of amino acids.

RNA A single-stranded nucleic acid that helps DNA create proteins in the cell.

somatic gene therapy The practice of modifying genes in somatic cells, which are body cells.

spectrophotometry A method used to measure the wavelength of light.

transgenic Describes an organism that contains foreign DNA.

virus An infectious organism that is composed of DNA and a coat of protein.

X-ray crystallography A method of taking X-ray photographs of crystal structures.

Further Information

BOOKS

Maddox, Brenda. *Rosalind Franklin: The Dark Lady of DNA*. New York, NY: Harper Collins, 2002.

Ridley, Matt, *Francis Crick: Discoverer of the Genetic Code*. New York, NY: Atlas Books, 2006.

Watson, James D. *The Secret of Life*. New York, NY: Alfred A. Knopf, 2003.

Wilkins, Maurice. *The Third Man of the Double Helix*. New York, NY: Oxford University Press, 2003.

WEBSITES

DNA from the Beginning
www.dnaftb.org/

A website from the Cold Spring Harbor Laboratory that details the history of the discovery of DNA through engaging interactives.

DNA Interactive
DNAi.org

A website that explains how DNA works and the history of DNA through interactives. Published by Cold Spring Harbor Laboratory, a renowned lab for the study of genetics.

The Rosalind Franklin Papers
profiles.nlm.nih.gov/ps/retrieve/Narrative/KR/p-nid/183

Browse through a collection of letters from Rosalind Franklin, curated by the National Library of Medicine.

VIDEOS

Amoeba Sisters
www.youtube.com/user/AmoebaSisters

A series of animated YouTube videos that explores different topics in biology (including DNA and RNA) in an engaging and accessible way.

Web of Stories—James Watson
www.webofstories.com/play/james.watson/1

An extended interview with James D. Watson about his childhood and his role in the discovery of the structure of DNA.

Bibliography

A&E Television Networks. "Rosalind Franklin Biography." Biography.com, 2016. Retrieved January 29, 2016 (http://www.biography.com/people/rosalind-franklin-9301344).

Aldridge, Susan. "The DNA Story." *Chemistry World*, April 2003. Retrieved January 29, 2016 (http://www.rsc.org/chemistryworld/Issues/2003/April/story.asp).

American Medical Association. "Gene Therapy." Retrieved January 29, 2016 (https://www.ama-assn.org/ama/pub/physician-resources/medical-science/genetics-molecular-medicine/current-topics/gene-therapy.page).

Bernal, J. D. "William Thomas Astbury." *Biographical Memoirs of Fellows of the Royal Society*, November 1963. Retrieved January 29, 2016 (http://www.leeds.ac.uk/heritage/Astbury/bibliography/Bernal_Astbury_obit.pdf).

Cambridge Physicists. "William Lawrence Bragg." Retrieved January 29, 2016 (http://www-outreach.phy.cam.ac.uk/camphy/physicists/bragg_prelim.htm).

The Cancer Genome Atlas. "Cancers Selected for Study." November 5, 2015. Retrieved January 29, 2016 (http://cancergenome.nih.gov/cancersselected).

The Cancer Genome Atlas. "Impact of Cancer Genomics on Precision Medicine for the Treatment of Cancer." Retrieved January 29, 2016 (http://cancergenome.nih.gov/cancergenomics/impact).

Carr, Steven M. "Levene's Tetranucleotide Hypothesis." MUN, 2015. Retrieved January 29, 2016 (https://www.mun.ca/biology/scarr/Tetranucleotide_Hypothesis.html).

Chemical Heritage Foundation. "Paul Berg, Herbert W. Boyer, and Stanley N. Cohen." Retrieved January 29, 2016 (http://www.chemheritage.org/discover/online-resources/chemistry-in-history/themes/pharmaceuticals/preserving-health-with-biotechnology/berg-boyer-cohen.aspx).

Clancy, Suzanne, and William Brown. "Translation: DNA to mRNA to Protein." Scitable, 2008. Retrieved January 29, 2016 (http://www.nature.com/scitable/topicpage/translation-dna-to-mrna-to-protein-393).

Collins, Francis S., and Harold Varmus. "A New Initiative on Precision Medicine." *New England Journal of Medicine*, February 26, 2015. Retrieved January 29, 2016 (http://www.nejm.org/doi/full/10.1056/NEJMp1500523).

Conova, Susan. "Columbia's Contribution to the DNA Structure: Chargaff's Rules." *P&S*, 2003. Retrieved January

29, 2016 (http://www.cumc.columbia.edu/psjournal/archive/fall-2003/dna.html).

Dahm, Ralf. "From Discovering to Understanding." NCBI, February 19, 2010. Retrieved January 29, 2016 (http://www.ncbi.nlm.nih.gov/pmc/articles/PMC2838690/).

DNA and Social Responsibility. "The Randall Letters: The DNA Story at King's Revisited." August 23, 2010. Retrieved January 29, 2016 (http://dnaandsocialresponsibility.blogspot.com/2010/08/randall-letters-dna-story-at-kings.html).

DNA from the Beginning. "The DNA Molecule Is Shaped Like a Twisted Ladder." Retrieved January 29, 2016 (http://www.dnaftb.org/19/bio.html).

Hall, Kersten. "William Astbury: Forgotten hero of DNA's discovery." *Guardian*, September 17, 2015. Retrieved January 29, 2016 (https://www.theguardian.com/science/2015/sep/17/william-astbury-forgotten-hero-of-dnas-discovery).

Harvard Medical School. "An Expanded Timeline of Personalized Medicine." 2016. Retrieved January 29, 2016 (https://hms.harvard.edu/news/harvard-medicine/expanded-timeline-personalized-medicine).

Highfield, Roger. "DNA: Three Letters That Spell Out a Discovery Made 50 Years Ago." *Telegraph*, February 28, 2003. Retrieved January 29, 2016 (http://www.telegraph.co.uk/news/science/science-news/3305922/DNA-three-letters-that-spell-out-a-discovery-made-50-years-ago.html).

Hunter, David J., Muin J. Khoury, and Jeffrey M. Drazen. "Letting the Genome Out of the Bottle—Will We Get Our Wish?" *New England Journal of Medicine*. Retrieved January 29, 2016 (http://www.nejm.org/doi/full/10.1056/NEJMp0708162).

International HapMap Project. "What Is the HapMap?" Retrieved January 29, 2016 (http://hapmap.ncbi.nlm.nih.gov/whatishapmap.html).

King's College London. "The Cell: Dissecting the New Anatomy." 2016. Retrieved January 29, 2016 (http://www.kcl.ac.uk/lsm/research/divisions/randall/about/history/cell.aspx#Randal).

Landrigan, Philip J., and Charles Benbrook. "GMOs, Herbicides, and Public Health." *New England Journal of Medicine*, August 20, 2015. Retrieved January 29, 2016 (http://www.nejm.org/doi/full/10.1056/NEJMp1505660).

Linus Pauling and the Race for DNA. "First Attempts." 2015. Retrieved January 29, 2016 (http://scarc.library.oregonstate.edu/coll/pauling/dna/narrative/page9.html).

Linus Pauling and the Race for DNA. "Watson and Crick." 2015. Retrieved January 29, 2016 (http://scarc.library.oregonstate.edu/coll/pauling/dna/narrative/page8.html).

Mackinnon, Alison, Inga Elgqvist-Saltzman, and Alison Prentice, eds. *Education into the 21st Century*. Bristol, PA: Falmer, Taylor & Francis, 1998.

Maddox, Brenda. "Mother of DNA." Newhumanist.org, 2002. Retrieved January 29, 2016 (https://newhumanist.org.uk/articles/532/mother-of-dna).

Maron, Dina Fine. "Big Precision Medicine Plan Raises Patient Privacy Concerns." *Scientific American*. Retrieved January 29, 2016 (http://www.scientificamerican.com/article/big-precision-medicine-plan-raises-patient-privacy-concerns/).

Mavilio, Fulvio, and Giuliana Ferrari. "Genetic Modification of Somatic Stem Cells." PMC, July 2008. Retrieved January 29, 2016 (http://www.ncbi.nlm.nih.gov/pmc/articles/PMC3327547/).

McFarling, Usha Lee. "Maurice Wilkins, 88; Had Role in Discovery of DNA." *Los Angeles Times*, October 7, 2004, 1-2. Retrieved January 29, 2016 (http://articles.latimes.com/2004/oct/07/local/me-wilkins7).

Mental Floss. "Rosalind Franklin and the Search for DNA." October 15, 2013. Retrieved January 29, 2016 (http://mentalfloss.com/article/53199/rosalind-franklin-and-search-dna).

NASA. "What Are Hard X-Rays?" Retrieved January 29, 2016 (http://hesperia.gsfc.nasa.gov/sftheory/xray.htm).

NCBI. "Safety of Genetically Engineered Foods: Approaches to Assessing Unintended Health Effects." 2004. Retrieved January 29, 2016 (http://www.ncbi.nlm.nih.gov/books/NBK215765/).

Oregon State University. "Linus Pauling and the Race for DNA:Watson and Crick." Last modified 2015. Retrieved January 29, 2016 (http://scarc.library.oregonstate.edu/coll/pauling/dna/narrative/page8.html).

The Pauling Blog. "Chargaff's Rules." August 25, 2009. Retrieved January 29, 2016 (https://paulingblog.wordpress.com/2009/08/25/the-dna-story-a-missed-opportunity/).

Plato. "Plato's TIMAEUS: Genesis of Other Animals." Elpenor's. Retrieved January 28, 2016 (http://www.ellopos.net/elpenor/physis/plato-timaeus/genesis-animals.asp).

PMC. "Molecular Basis for the Herbicide Resistance of Roundup Ready Cops." August 17, 2006. Retrieved January 29, 2016 (https://www.ncbi.nlm.nih.gov/pmc/articles/PMC1559744/).

Pollack, Andrew. "Genetically Engineered Salmon Approved for Consumption." *New York Times*, November 19, 2015. Retrieved January 29, 2016 (http://www.nytimes.com/2015/11/20/business/genetically-engineered-salmon-approved-for-consumption.html?_r=1).

Profiles in Science. "The Linus Pauling Papers." Retrieved January 29, 2016 (https://profiles.nlm.nih.gov/ps/retrieve/Narrative/MM/p-nid/54).

Profiles in Science. "The Rosalind Franklin Papers." Retrieved January 29, 2016 (https://profiles.nlm.nih.gov/ps/retrieve/Narrative/KR/p-nid/183).

Rzepa, Henry. "A Short History of Molecular Modelling: 1860–1890." Chemistry With a Twist. Retrieved January 29, 2016 (http://www.ch.ic.ac.uk/rzepa/blog/?p=3472).

Schrodinger, Erwin. "What Is Life?" Stanford, 1944. Retrieved January 29, 2016 (http://whatislife.stanford.edu/LoCo_files/What-is-Life.pdf).

Scitable. "Maurice Wilkins: Behind the Scenes of DNA." 2014. Retrieved January 29, 2016 (http://www.nature.com/scitable/topicpage/maurice-wilkins-behind-the-scenes-of-dna-6540179).

Smithsonian National Museum of Natural History. "What Does It Mean to Be Human?" Last modified January 28, 2016. Retireved January 29, 2016 (http://humanorigins.si.edu/evidence/genetics).

Stein, Rob. "Scientists Debate How Far to Go in Editing Human Genes." NPR, December 3, 2015. Retrieved January 29, 2016 (http://www.npr.org/sections/health-shots/2015/12/03/458212497/scientists-debate-how-far-to-go-in-editing-human-genes).

Su, Pascal. "Direct-to-Consumer Genetic Testing: A Comprehensive View." PMC, September 20, 2013. Retrieved January 29, 2016 (http://www.ncbi.nlm.nih.gov/pmc/articles/PMC3767220/).

Summers, William C. "How Bacteriophage Came to Be Used by the Phage Group." *Journal of the History of*

Biology. Retrieved January 29, 2016 (http://www.jstor.org/stable/4331263?seq=1#page_scan_tab_contents).

Wainwright, Martin. "Sidelined Scientist Who Came Close to Discovering DNA Is Celebrated at Last." *Guardian*. Last modified November 23, 2010. Retrieved January 29, 2016 (https://www.theguardian.com/science/2010/nov/23/william-astbury-dna-scientist).

Index

Page numbers in **boldface** are illustrations. Entries in **boldface** are glossary terms.

About the Author

Jenny J. Chen is an award-winning science journalist. Her work has appeared in the *Atlantic*, NYTimes.com, and *New York Magazine*, and on NPR. In 2014 and 2015, she was awarded a PRX STEM grant to produce stories for NPR member stations across the country. In 2014, she received a grant from the DC Humanities Council to produce a radio documentary series on growing up mixed race in Washington, DC. Jenny has also received numerous fellowships and awards to cover health, aging, minority issues, and climate change. She has spoken about journalism and the role of ethnic media at the Smithsonian Folklife festival. In another life, she also had a play produced at Arena Stage and the Kennedy Center.

She currently lives in Washington, DC, with her family. Her website is www.jennyjchen.com.